Studies in Computational Intelligence

Volume 783

Series editor

Janusz Kacprzyk, Polish Academy of Sciences, Warsaw, Poland
e-mail: kacprzyk@ibspan.waw.pl

The series "Studies in Computational Intelligence" (SCI) publishes new developments and advances in the various areas of computational intelligence—quickly and with a high quality. The intent is to cover the theory, applications, and design methods of computational intelligence, as embedded in the fields of engineering, computer science, physics and life sciences, as well as the methodologies behind them. The series contains monographs, lecture notes and edited volumes in computational intelligence spanning the areas of neural networks, connectionist systems, genetic algorithms, evolutionary computation, artificial intelligence, cellular automata, self-organizing systems, soft computing, fuzzy systems, and hybrid intelligent systems. Of particular value to both the contributors and the readership are the short publication timeframe and the world-wide distribution, which enable both wide and rapid dissemination of research output.

More information about this series at http://www.springer.com/series/7092

Lyndon White • Roberto Togneri
Wei Liu • Mohammed Bennamoun

Neural Representations of Natural Language

 Springer

Lyndon White
Department of Electrical, Electronic and
    Computer Engineering, School of
    Engineering, Faculty of Engineering and
    Mathematical Sciences
The University of Western Australia
Perth, WA, Australia

Roberto Togneri
Department of Electrical, Electronic and
    Computer Engineering, School of
    Engineering, Faculty of Engineering and
    Mathematical Sciences
The University of Western Australia
Perth, WA, Australia

Wei Liu
Department of Computer Science and
    Software Engineering, School of Physics,
    Mathematics and Computing, Faculty of
    Engineering and Mathematical Sciences
The University of Western Australia
Perth, WA, Australia

Mohammed Bennamoun
Department of Computer Science and
    Software Engineering, School of Physics,
    Mathematics and Computing, Faculty of
    Engineering and Mathematical Sciences
The University of Western Australia
Perth, WA, Australia

ISSN 1860-949X   ISSN 1860-9503   (electronic)
Studies in Computational Intelligence
ISBN 978-981-13-4320-9   ISBN 978-981-13-0062-2   (eBook)
https://doi.org/10.1007/978-981-13-0062-2

This Springer imprint is published by the registered company Springer Nature Singapore Pte Ltd.
The registered company address is: 152 Beach Road, #21-01/04 Gateway East, Singapore 189721, Singapore

Preface

We do, therefore, offer you our practical judgements wherever we can. As you gain experience, you will form your own opinion of how reliable our advice is. Be assured that it is not perfect!

Press, Teukolsky, Vetting and Flannery, *Numerical Recipes*, 3rd ed., 2007

Some might wonder why one would be concerned with finding good representations for natural language. To answer that simply, all problem-solving is significantly simplified by finding a good representation, whether that is reducing a

---

A wealth of other materials

Wikipedia articles on ML and NLP tend to be of reasonable quality, Coursera offers several courses on the topics, Cross Validated Stack Exchange has thousands of QAs, and the documentation for many machine learning libraries often contains high-quality tutorials. Finally, preprints of the majority of the papers in the fields are available on arXiv.

---

real-world problem into a system of inequalities to be solved by a constrained optimizer or using domain-specific jargon to communicate with experts in the area. There is a truly immense amount of existing information, designed for consumption by humans. Much of it is in text form. People prefer to create and consume information in natural language (such as prose) format, rather than in some more computationally accessible format (such as a directed graph). For example, doctors prefer to dictate clinical notes over filling in forms for a database.

One of the core attractions of a deep learning system is that it functions by learning increasingly abstract representations of inputs. These abstract representations include useful features for the task. Importantly, they also include features applicable even for more distantly related tasks.

It is the general assumption of this book that it is not being read in isolation. That the reader is not bereft of all other sources of knowledge. We assume not only can the reader access the primary sources for the papers we cite, but also that they are able to discover and access other suitable materials, in order to go deeper on related areas. There exist a wealth of blog posts, video tutorials, encyclopaedic article, etc., on machine learning and on the mathematics involved.

In general, we do assume the reader is familiar with matrix multiplication. In general, we will define networks in terms of matrices (rather than sums), as this is more concise, and better reflects real-world applications, in terms of code that a programmer would write. We also assume a basic knowledge of probability, and an even more basic knowledge of English linguistics. Though they are not the intended audience, very little of the book should be beyond someone with a high-school education.

The core focus of this book is to summarize the techniques that have been developed over the past 1–2 decades. In doing so, we detail the works of several dozen papers (and cite well over a hundred). Significant effort has gone into describing these works clearly and consistently. In doing so, the notation used does not normally line up with the original notations used in those papers; however, we ensure it is always mathematically equivalent. For some techniques, the explanation is short and simple. For other more challenging ideas, our explanation may be much longer than the original very concise formulation that is common in some papers.

For brevity, we have had to limit discussion of some aspect of natural language. In particular, we have neglected all discussion on the notion that a word may be made up of multiple tokens, for example, made up of. Phrases do receive some discussion in Chap. 6. Works such as Yin and Schütze (2014) exploration deserve more attention than we have space to give them.

Similarly, we do not have space to cover character-based models, mainly character RNNs, and other deep character models such as Zhang and LeCun (2015). These models have relevance both as sources from which word representations can be derived (Bojanowski et al. 2017), but more generally can be used for end-to-end systems. Using a purely character-based approach forces the model to learn tokenizing, parsing, and any other feature extraction internally. We focus instead on models that work within larger systems that accomplish such preprocessing.

Finally, we omit all discussion of attention models in recurrent neural networks (Bahdanau et al. 2014). The principles to the application remain similar to using normal recurrent neural networks in the applications discussed in this book. The additional attention features allow several improvements, making it the state of the art for many sequential tasks. We trust the reader with the experience of such models and will see how they may be applied to the uses of RNNs discussed here.

This book makes extensive use of text in grey boxes. This is used to provide reminders of definitions and comments on potentially unfamiliar notations. As well as to highlight non-technical aspects (such as societal concerns), and conversely overly technical aspects (such as implementation details) of the works discussed. They are also used to give the titles to citations (to save flicking to the huge list of references at the back of the book) and for the captions to the figures and more generally to provide non-essential but potentially helpful information without interrupting the flow of the main text. One could read the whole book without ever reading any of the text in grey boxes, as they are not required to understand the main text. However, one would be missing out on a portion of, some very interesting, content.

Perth, Australia

Lyndon White
Roberto Togneri
Wei Liu
Mohammed Bennamoun

References

Bahdanau, Dzmitry, Kyunghyun Cho, and Yoshua Bengio. 2014. Neural machine translation by jointly learning to align and translate. In *CoRR* abs/1409.0473. arXiv:1409.0473.

Bojanowski, Piotr, Edouard Grave, Armand Joulin, and Tomas Mikolov. 2017. Enriching word vectors with subword information. *Transactions of the Association for Computational Linguistics*, vol. 5, 135–146.

Yin, Wenpeng and Hinrich Schütze. 2014. An exploration of embeddings for generalized phrases. In *ACL* 2014, p. 41.

Zhang, Xiang and Yann Le Cun. 2015. Text understanding from scratch. In *CoRR proceedings of the 52nd annual meeting of the association for computational linguistics*, vol. 11.

Contents

Notations

The following notation is used throughout this work.

a	A scalar (real, integer, or word/token)
$\tilde{a}$	A vector, nominally a column vector
A	A matrix
$\mathcal{A}$	A sequence, including a data set or a sequence of words
$\mathbb{V}$	A set, e.g. the vocabulary
$\tilde{x}_i$	The ith element of the vector $\tilde{x}$
$X_{i,j}$	The row i and column j element X
X_i	The ith column *vector* of the matrix X
$X_{i,:}$	The ith row *vector* of the matrix X
w^t	A scalar tth element of some sequence
W^f	A matrix disambiguated by the name f
$[A\ B]$	The horizontal concatenation of A and B
$[A;B]$	The vertical concatenation of A and B
$P(\ldots)$	A probability (estimated or ground truth)
A	A random variable (when not a matrix)
$\hat{y}$	A network output value, corresponding to target value y or $\tilde{y}$ a vector or scalar quantity as appropriate

Words are treated as integers
We consistently notate words, as if they were scalar integer values. writing, for example, w^1 as to be the first word in a sequence, which is then used as an index: $C_{:,w^i}$ is its corresponding word vector, from the embedding matrix C.

Superscripts and Subscripts

Readers may wonder why we are using x_i, and x^i. Would not x_i suffice? Why differentiate between elements of a sequence and elements of a vector?

The particularly problematic case is that we often want to represent taking the ith element of a vector that is the tth element of a sequence of vectors. The vector, we would call $\tilde{x}^t$, its i element is $\tilde{x}_i^t$.

This is also not ambiguous with the matrix indexing notation $X_{i,t}$.

Rarely, a superscript will be actually an exponent, e.g. $x^{\frac{2}{3}}$. This should be apparent when in this case. Far more common is the natural exponent which we write $\exp(x)$.

*Whether one sees the artificial neural network technique
described below as learning or as optimization depends largely
on one's background and one's theoretical likes and dislikes. I
will freely use "learning" in the remainder of this section
because that term is traditionally used in the neural networks
literature, but the reader should feel free to substitute
"optimization" if (s)he finds the other term offensive. Please
contact the author for an Emacs lisp function to enforce the
usage of choice*

J. M. G. Lammens, PhD Dissertation, A Computational Model
of Color Perception and Color Naming, State University of New
York, 1994

Abstract

This chapter covers the crucial machine learning techniques required to understand the remained of the book: namely neural networks. Readers already familiar
with neural networks can freely skip this chapter. Readers interested in a more
comprehensive coverage of all aspects of machine learning are referred to the
many textbooks on this subject matter.

Suggested Reading
As this is not a comprehensive introduction we recommend the reader to look
elsewhere for additional reading. In particular we recommend the free online
web-book: Nielsen and Michael (2015), and the comprehensive: Goodfellow
et al. (2016).

Machine learning, particularly neural networks, has become a hot topic in recent
years. The core notion of machine learning is to learn to perform a function based
on examples. This is in-contrast to "regular programming" where code is written to
accomplish the function based on the programmer's analysis of the task. Machine
learning is at its most useful when it is difficult to articulate explicit rules for finding

an output from a given input; but for which there are many examples of such. For example, the rule of English spelling "I before E, except after C", is known to be often incorrect. It is hard to write a rule about letter order. However, a dictionary (or any other corpus) will have thousands of words showing the correct order. By applying suitable machine learning techniques, a system could learn to determine the correct spelling of words containing ie and ei, and inform the users as to if an input word is correct, even if the word never occurred in the training dictionary. This is because the machine learning algorithm (ideally) discovers generalisable rules, from the training examples. For most machine learning methods, including neural networks, these rules are not in a readily interpretable form, but are stored as numerical parameters of the model. It is for this reason they are often called black-box models.

1.1   Neural Networks

Multilayer perceptron or Artificial Neural Network?
An artificial neural network is sometimes called a multilayer perceptron. This is in reference to the (distantly) related perceptron learning algorithm. This term has generally fallen out of favour in newer works, with Geoffrey Hinton, who originally coined the term, expressing his own regret at the naming. Sometimes the term multilayer perceptron is used to distinguish feed-forward neural networks, from other neural network related technologies, such as restricted Boltzmann machines, Hopfield networks, self-organizing maps etc. Unless otherwise specified, we use the term neural network to refer to a feed-forward network as discussed in this chapter.

A neural network is one particular family of machine learning algorithms. It is important to understand that neural networks are not emulations of the brain, they are algorithms that were inspired by the ideas about how the brain worked (Hebb 1949). Some advancements have also been inspired by the functioning of the brain, but many others have come from statistical methods or elsewhere. A neural network is no more similar to a brain, than it is to a Fourier transform.

The core idea behind a neural network is to represent the transformation of the input to the output as links between neurons in a sequence of layers. This is shown in Fig. 1.1. Each neuron has a weighted connection to the neurons in the layer below, and it applies a nonlinear function to this weighted sum, to determine its own output. The output is connected to the inputs in the layer above, or is the final output.

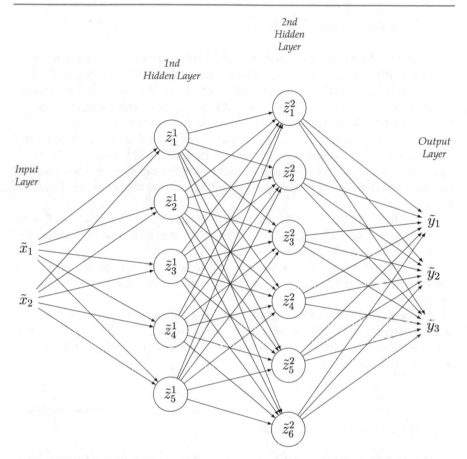

Fig. 1.1 A diagram of a neural network, with 2 inputs, and 3 outputs, and 2 hidden layers of width 5 and 6 neurons respectively

Don't implement neurons
Typical object orientated programming teaches one to look for objects. A obvious object in a neural network would be a neuron. Another object could be the connection (weight). Perhaps a different subclass of neuron would be for each activation function. Do not do this. It will be incredibly slow in almost any programming languages – where objects are always stored by reference, thus requiring a pointer dereferencing for each operation on each neuron. Object orientated is not a good paradigm for this kind of technical computing. Instead program in terms of matrix's and vectors. Potentially using objects for layers, or for the entire network. This allows one to take advantage of efficient matrix algebra routines such as in BLAS and LAPACK.

1.2 Function Approximation

The Universal Approximation Theorem (UAT) tells us that a neural network can be used to approximate almost any function (Mhaskar et al. 1992; Leshno et al. 1993; Sonoda and Murata 2017). Here a function should be understood in the general sense – classification is a function from observations to classes, regression is a function from observed values to target values. More precisely, for any continuous bounded function, the UAT states that there exists a neural network that can approximate that function, to any given degree of accuracy. Further, it states that the network only needs a single hidden layer. In a less rigorous sense, of approximation, such a continuous bounded function can approximate any function, over a restricted sub-domain of input values.

The earliest proofs of this was restricted to bounded activation functions like the sigmoid. More recent work has extended this for unbounded activation functions like ReLU (Sonoda and Murata 2017).

A network without any hidden layers is only able to approximate linear functions. It is equivalent to a perceptron, or a linear SVM (Rumelhart et al. 1986; Minsky et al. 2017).

However, the UAT does not tell us the values for those weight parameters, nor does it promise that any method of training will ever reach them in finite time. More significantly it does not inform as to how wide *sufficiently wide* is, nor if deeper is more efficient than wider.

1.3 Network Hyper-Parameters

Hyper-parameters versus Parameters
The weights and biases of the network are called its parameters. They are optimised during training. The other features of the network, including its topology, activation functions, and the training method (which often has its own set of options), are called the hyper-parameters.

The key determination in applying neural networks to any problem, is the choice of the hyper-parameters. Including, the number of hidden layers, and their widths. Neural networks can largely be thought of as black-boxes, that can be trained to accomplish tasks given a set of training examples. Beyond recognising how to express the problem, the choice of hyper-parameters is the most important decision when employing a neural network based solution.

1.3.1   Width and Depth

What happened to Unsupervised Pretraining?
Hinton et al. (2006), Bengio et al. (2007) lead to a new resurgence of interests in neural networks. Pretraining with Deep Belief Networks allowed for deep networks to be trained. It was held for many years that it was necessary to use layer-wise unsupervised pretraining, even when one has only supervised data. Glorot et al. (2011) showed that with ReLU units, a deep network could be trained directly and achieve the same performance. As such unsupervised pretraining is now a more niche technique used primarily when there is an excess of unsupervised data.

While the UAT has shown that a single hidden layer is sufficient, it is not necessary optimal in terms of achieving best performance in practice. For many problems it is better to have a deeper rather than wider network. This realisation and the techniques to overcome issues relating to deeper networks lead the current deep learning trend.

In the last decade, deep nets have come back into fashion. In brief the reasoning for this is a combination of better techniques (e.g. ReLU, unsupervised pretraining), better hardware (e.g. GPUs and Xeon Phi), and more labelled data (e.g. from crowd-sourcing platforms such as Amazon Mechanical Turk.) This has allowed problems that were previously considered unsolvable with earlier shallow networks to be solved with deep networks achieving state of the art results.

The choice of the number of hidden layers (depth), and the sizes (width) is a key parameter in designing a neural network. It is arguably *the* key parameter. It can be assessed using a hyper-parameter sweep on a validation or development dataset. It is a particularly relevant parameter for our purposes, as hidden layers normally are directly linked to our representations of natural language.

1.3.2   Activation Functions

Notation: Generic Activation Function φ:
Though-out this book, when the activation function is not significant to the problem at hand, we will represent it with the symbol $\varphi(\cdot)$.

Activation functions summary:
identity:	range $(-\infty, \infty)$
sigmoid:	range $(0, 1)$
softmax:	range $(0, 1)$
tanh:	range $(-1, 1)$
ReLU:	range $(0, \infty)$
ReLU6:	range $(0, 6)$

The universal approximation theorem places a number of requirements on the choice of activation function. These requirements have been progressively relaxed by works such as Leshno et al. (1993), Sonoda and Murata (2017). This section discusses the activation functions in common use today.

The choice of activation function determines the range of output of a neuron. For the hidden layers these generally perform relatively similarly. Sigmoid, tanh, and ReLU are the most common activation functions used for hidden layers. The range of output of a hidden layer does not normally matter, as it is by definition hidden. More significantly, the final activation function or the output function does always matter. The output function should be chosen for the task at hand. As will be discussed in Chap. 2 this also has significance when designing gated neural network structures.

Beyond determining the range of the output, its derivative also determines the behaviour during training via gradient descent based training methods. This is significant for activation functions like ReLU and ReLU6 discussed in the following sections.

The activation functions, other than the softmax layer, are applied element-wise to vector inputs. As such the scalar definitions are given here. We write $y = \varphi(z)$, for y being the output, and z being the weighted input from the layer below. To represent the whole layer including weights and biases we write:

$$\tilde{y} = \varphi(z) = \varphi(W\tilde{x} + \tilde{b}) \tag{1.1}$$

1.3.2.1 Identity (Affine/Linear)

The most basic activation function is the identity. To simply output the weighted sum of the inputs, without applying a non-linearity. It is commonly used as an output function for regression tasks, as it imposes no (additional) bounds.

$$\tilde{y} = \tilde{z} \tag{1.2}$$

Often the use of the identity activation function will be described in terms of the whole layer, including the weight and bias. The layer is often called an affine layer as it corresponds to an affine transformation of its input. When the bias is fixed at zero, this is a linear transformation of the input.

$$\tilde{y} = W\tilde{x} + \tilde{b} \tag{1.3}$$

In this equation W and $\tilde{b}$ are a trained weight matrix and bias vector respectively, and $\tilde{x}$ is the output of the layer below.

Using an affine output layer on-top of the hidden layers allows the learned scaling and shifting of any of the other activation functions. However, when this is not needed it increases the difficulty of training for no benefit.

Identity functions cannot be used as hidden layer activation functions. A stack of affine layers simplifies to be equivalent to a single affine layer, which means that

Fig. 1.2 The sigmoid
function

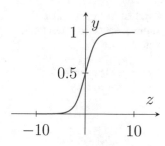

the network can only approximate linear functions. A non-linearity is required, as is
provided by most activation functions.

1.3.2.2 Sigmoid

Sigmoid is the classic neural network activation function. It is shown in Fig. 1.2.

$$y = \sigma(z) = \frac{1}{1 + \exp(-z)} \tag{1.4}$$

Sigmoid outputs values between zero and one. If the network has a single output then
this is a fuzzy boolean. It can be used to classify into two categories. It is sometimes
referred to the logistic output function as it forms the basis of logistic regression.
(Conversely, sometimes the word sigmoid is used to refer to all functions of such an
S shape.)

Note that this function has the property that, $\sigma(-z) = 1 - \sigma(z)$. This means
negating the input to the final sigmoid in a binary classification network, is the same
as switching which class is considered as true.

Multinomial logistic regression

A neural network classifier with a softmax output layer is sometimes called a
multinomial logistic regression network (especially if there are no hidden layers), or even just a logistic regression network. This can lead to some confusion
with a sigmoid outputting network. However, the similarity can be understood
by considering a softmax output layer with 2 classes. For $\tilde{z} = [\tilde{z}_1, \tilde{z}_2]$ defining
$v = \tilde{z}_1 - \tilde{z}_2$

A (binary) logistic classification using the same inputs can be written:
$P(Y = 1) = \sigma(v) = \frac{1}{1+\exp(\tilde{z}_2 - \tilde{z}_1)}$.
The 2 class softmax formulation is:
$\text{smax}(\tilde{z})$
$$= \left[\frac{\exp(\tilde{z}_1)}{\exp(\tilde{z}_1) + \exp(\tilde{z}_2)}, \frac{\exp(\tilde{z}_2)}{\exp(\tilde{z}_1) + \exp(\tilde{z}_2)} \right]$$
$$= \left[\frac{1}{1+\exp(\tilde{z}_2 - \tilde{z}_1)}, \frac{1}{\exp(\tilde{z}_1 - \tilde{z}_2) + 1} \right]$$
$$= [\sigma(v), \sigma(-v)]$$
$$= [\sigma(v), 1 - \sigma(v)]$$
$$= [P(Y = 1), P(Y = 2)]$$

Fig. 1.3 An example of a
softmax activation in use for
with 3 classes for
$z \mapsto \mathrm{smax}([z; \frac{z}{10}; -2z)]$.
Notice at all points the 3
outputs sum to 1

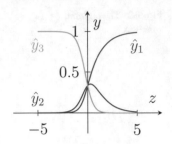

1.3.2.3 Softmax

The standard output activation function for multiclass classification is softmax. Formally, training a so-called softmax classifier is regression to a class probability, rather than a true classifier as it does not return the class but rather a prediction of each class's likelihood. We will use the terms interchangeably.

Defining the layer $\tilde{y}$ in terms of its elements, recall that by our notation convention $\tilde{y}_i$ is the ith element of the vector $\tilde{y} = \begin{bmatrix} \tilde{y}_1, \tilde{y}_2, \dots \tilde{y}_n \end{bmatrix}$:

$$\tilde{y}_i = (\mathrm{smax}(\tilde{z}))_i = \frac{\exp(\tilde{z}_i)}{\sum_{j=1}^{j=N} \exp(\tilde{z}_j)} \tag{1.5}$$

$\tilde{y}$ is a vector of length n, where n the number of classes in the classification.

These elements have values between zero and one, and sum to one. For example as shown in Fig. 1.3. They thus define a discrete probability mass function.

When used for classification $\tilde{y}_i$ gives the probability of being in class i.

$$P(Y = i \mid \tilde{z}) = (\mathrm{smax}(\tilde{z}))_i = \tilde{y}_i \tag{1.6}$$

Softmax is a *soft*, i.e. differentiable, version of what could be called *hard-max*. In this conceptual hard-max function, the largest output is set to one, and the others set to zero. It is non-differentiable and thus not suitable for use in a neural network. In softmax the larger values is proportionally increased to be closest to one, and the smaller to be closest to zero.

Further consideration of the softmax is given in Sect. 2.5.1.1.

1.3.2.4 Tanh

The hyperbolic tangent function is a scaled and shifted sigmoid function. Such that the outputs are restricted to be between -1 and $+1$. It is shown in Fig. 1.4.

$$y = \tanh(z) = \frac{\exp(2z) - 1}{\exp(2z) + 1} = 2\sigma(2z) + 1 \tag{1.7}$$

Fig. 1.4 The tanh function

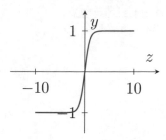

Fig. 1.5 The ReLU function

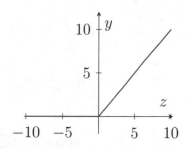

1.3.2.5   ReLU
The Rectified Linear Unit (ReLU) is a more recently popularized activation function (Dahl et al. 2013). It comes from earlier neuroscience work (Hahnloser 1988). It is shown in Fig. 1.5.

$$y = ReLU(z) = \max{(0, z)} = \begin{cases} z & 0 \leq z \\ 0 & z < 0 \end{cases} \tag{1.8}$$

Values are restricted to be at non-negative. As this function has derivative zero for $z < 0$, once a unit is turned off, it is not turned back-on. During gradient descent (see Sect. 1.4.0.1) the derivative is used to modify the weights, as it is zero it the weights will never change once turned off. This is commonly called the neuron dying. Further to this, using (zero mean) random initialization ensures that a significant portion of all neurons will initially be dead. This results in sparse connections between the layers. This has been found to be a good thing (Glorot et al. 2011). ReLU is very commonly used as a hidden layer activation function for deep networks. It also helps with the gradient vanishing and gradient exploding problems that occur in deep networks (Glorot et al. 2011).

1.3.2.6   RELU6
Closely related to ReLU is ReLU6 (Krizhevsky et al. 2010). This is another linear unit the saturates at 0, but also at 6 (Fig. 1.6).

Fig. 1.6 The RELU6
function

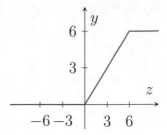

$$y = ReLU6(z) = \max\left(0, \min\left(6, z\right)\right) = \begin{cases} 6 & 6 < z \\ z & 0 \leq z \leq 6 \\ 0 & z < 0 \end{cases} \qquad (1.9)$$

Why are we bringing up ReLU6?
ReLU6 is one of the more obscure activation functions. As such it may seem a bizarre inclusion in this shorter introduction. However, we have found it surprisingly successful as an activation function, e.g. in the auto-encoder example which follows in Sect. 1.5.2. It also nicely illustrates the flexibility and indeed arbitrariness in how activation functions can be defined.

This has similar advantages to ReLU. However, as well as units being able to die to zero, it can also die to positive. The bounding at 6, rather than any other number is rather arbitrary – particularly given scaling with the intervening affine layers. The important point is that it is bound both above and below. This greatly simplifies any proofs of theoretical capabilities of various networks.

1.3.2.7 Other Activation Functions
There are numerous other activation functions. Most are of interest for use in hidden layers such as Leaky RELU, Softplus, ELU and many others. For specialized tasks, other functions might be used, such as atan2 (White 2016).

1.4 Training

The process of training the network is the process of solving a very high dimensional, nonlinear, non-convex global optimization problem.

> **Evaluation function versus Loss function**
>
> We must distinguished between two types of *error function*. The *evaluation function* is the metric we are truly trying to improve, this could be accuracy for classification, BLEU score for translation, F1 score for retrieval etc. It is applied to the whole system (which may be greater than just the neural network), and the system may be evaluated in different ways using different evaluations. As the evaluation is often not differentiable, a proxy for it which we call the **loss function** is used in training. For example, squared error for regression, or cross-entropy for classification. The loss function is applied to the output of the network during training to calculate the error between the network output for a given input, and the target output from the training example. The loss function is designed such that minimising the loss function also (ideally) results in the true evaluation function being optimised.

Finding the globally optimal solution to such a function is itself a very difficult problem. Nonlinear and non-convex problems can have local minima that are not global optima. This means they cannot guarantee to reach a global optima by gradient descent. However, this does not pose an issue for most neural networks as finding the true global optima is not required, merely finding a *good enough* set of weights to get a *good enough* result.

To train the network one must find the values for the networks weight and bias parameters, such that the total loss function is minimized over the training set.

To define the training problems as an optimisation problem one first considers the loss function for a single training case. For a target output y and an actual output $\hat{y}$, (considering the scalar case) a loss function is defined $loss(y, \hat{y})$. For example the squared error loss (sometimes called L2 loss, though that risks confusion with L2 weight penalty for regularisation) used in regression is defined by

$$SE(y, \hat{y}) = (y - \hat{y})^2 \tag{1.10}$$

The cross-entropy loss used in binary classification is defined by

$$CE(y, \hat{y}) = -\big((1 - y)\log(1 - \hat{y}) + y\log(\hat{y})\big) \tag{1.11}$$

The choice of loss function depends on the purpose of the network.

When network (represented by the function f) is composed into the loss function the per training case loss is given by $loss(\tilde{y}, \tilde{f}(\tilde{x}, \theta))$. $\tilde{f}(\tilde{x}, \theta)$ is the function that executes some neural network, with weight and bias parameters all grouped into θ, upon an input $\tilde{x}$ where the target output is $\tilde{y}$.

The total loss ($Loss$) is defined by taking the sum over the whole training set ($\mathcal{X}$) of input output pairs.

$$Loss(\theta, \mathcal{X}) = \sum_{\forall(\tilde{x}, \tilde{y}) \in \mathcal{X}} loss(\tilde{y}, \tilde{f}(\tilde{x}, \theta)) \tag{1.12}$$

Equation (1.12) nothing but a nonlinear function to be minimised by adjusting the values of the weights and biases in θ.

This can be given to off-the-shelf unconstrained nonlinear optimisation algorithms (Ngiam et al. 2011) such as L-BFGS (Nocedal and Jorge 1980). Alternatively, and more commonly, the loss and updated can be processed iteratively on sub-sets, called minibatches, of the training data at a time. The extreme case of this is to update for every single training example – this is called online training. Updating on minibatches is generally preferred over updating per example as it results in less small "noisy" weight fluctuations which may prevent reaching local minima. It is preferred over full-batch (on the whole training set) as the more frequent updates allow for faster learning (LeCun et al. 2012). Minibatch optimisation is often performed using algorithms specifically targeted at machine learning applications such as Adam (Kingma and Ba 2014). In either case, almost all optimizers used for neural networks rely on gradient descent.

AdaGrad, AdaDelta, and Adam
AdaGrad (Duchi et al. 2011), AdaDelta (Zeiler 2012) and Adam (Kingma and Ba 2014) are effectively successive versions of an optimisation algorithm, specifically targeted at the machine learning case. These algorithms dynamically adapt the learning rate during training. There are several other algorithms in the family, for example RMSprop. In recent works, Adam is the most commonly used optimiser. Dauphin et al. (2015) presents another algorithm from this family, and considers how the adaptive learning rate is related to the second order derivative, which is the fundamental innovation in more traditional quasi-Newtonian methods (over first-order methods like plain gradient descent.).

1.4.1 Gradient Descent and Back-Propagation

Back-propagation versus Gradient Descent?
Formally speaking, back-propagation finds the gradients, and gradient descent uses those gradients to update the network parameters. However, the terms are often used interchangeably.

Gradient descent is the basis of most commonly used optimisers in machine learning. In gradient descent based methods the derivatives of the loss function relative to the parameters are used to update those parameters. The parameters in this case are the network weights and biases. To calculate the gradients the back-propagation algorithm is used.

Back-propagation is simply a method of applying the well-known chain-rule of calculus to the neural network loss functions (Rumelhart et al. 1986). It allows one to calculate: $\frac{\partial Loss}{\partial W_{i,j}}$ for any weight (or bias) parameter $W_{i,j}$ in the network. These partial

Fig. 1.7 A visualisation of a possible error (loss) surface. The arrow indicates a possible step from one set of parameters (weights) down to another set of parameters with lower loss, via gradient descent

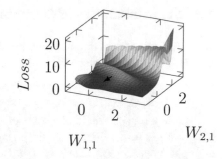

derivatives can always be found as the overall loss function, including the network, is differentiable[1] (when using typical activation functions and loss functions).

Using these gradients the network parameters can thus be updated. Updating the parameters to a new value in the direction of the gradient decreases the loss (for the right step size). This means for each $W_{i,j} \in \theta$ updating its value according to the local gradient.

$$W_{i,j} \leftarrow W_{i,j} - \alpha \frac{\partial Loss}{\partial W_{i,j}} \tag{1.13}$$

where α is an update step size, commonly called the learning rate. The more advanced methods discussed earlier, such as Adam, and L-BFGS, are extensions along this core principle.

If one considers a plot with the loss given on the vertical axis, and the values of the parameters as describing a position on the horizontal axes, then gradient descent is moving that the parameter-point downward on that surface based on local estimate of the slope. Such a plot is shown in Fig. 1.7 Most real networks have far more than 2 parameters, of-course, however the process of gradient decent remains the same.

These examples are available online
For a practical introduction to the networks discussed here, worked examples of these network are available in the accompanying blog post at http://white.ucc.asn.au/NNforNLPBook/NNexamples. They are written in the Julia programming language, making use of the TensorFlow.jl package.

---

[1]Technically most networks are differentiable almost everywhere. For some functions like ReLU they have some single points of where the derivative is not defined. But the derivatives here can be forced to a known value.

1.5   Some Examples of Common Neural Network Architectures

We will discuss more linguistically relevant neural networks in the following chapters. However, to introduce the topic, we will give some basic examples.

> **FashionMNIST**
> Is a new image classification benchmarking dataset (Xiao et al. 2017). It is a task to classify images of shirts, shoes and bags etc. It is an exact drop-in replacement for MNIST. It was created to allow the same benchmarking scripts to be used; while using a different data set. This combats extremely high use of MNIST leading to implicit test set information leakage in the design of ML systems (where systems are base on systems which perform well on MNIST due to more of such systems being published.); as well as being a more difficult task, thus allowing better discrimination between systems.

1.5.1   Classifier

A classifier on the MNIST challenge is one of the most common introductory networks. The MNIST dataset is a collection of images of hand written digits, which much be classified as to which digit they are.

A basic network to complete this task is is shown in Fig. 1.8.

The network is defined by the equations:

$$\tilde{z} = \varphi(W^1 x + \tilde{b}_1) \tag{1.14}$$

$$\hat{y} = \text{smax}(W^2 z + \tilde{b}_2) \tag{1.15}$$

$$P(Y = i | X = \tilde{x}) = \hat{y}_i \tag{1.16}$$

$$= \left[\text{smax}(W^2 \varphi(W^1 \tilde{x} + \tilde{b}^1) + \tilde{b}^2) \right]_i \tag{1.17}$$

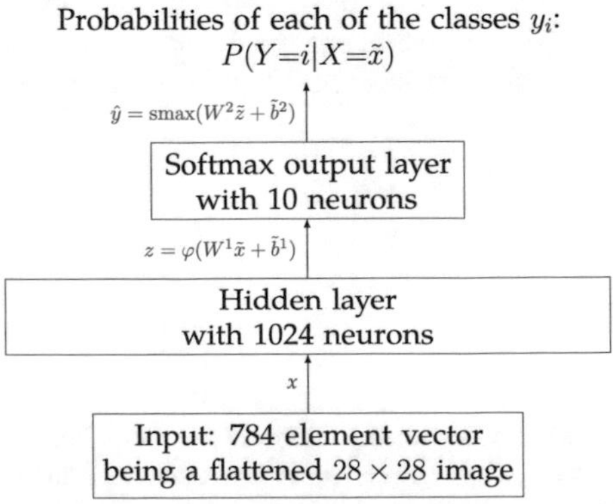

Probabilities of each of the classes y_i:
$$P(Y{=}i | X{=}\tilde{x})$$

$\hat{y} = \text{smax}(W^2\tilde{z} + \tilde{b}^2)$ ↑

Softmax output layer with 10 neurons

$z = \varphi(W^1\tilde{x} + \tilde{b}^1)$ ↑

Hidden layer with 1024 neurons

x ↑

Input: 784 element vector being a flattened 28×28 image

Fig. 1.8 The structure of a simple classifier network for MNIST

for X the input variable of grey-scale pixels intensities, and Y the output as a class. The output is represented with 1 hot encoding, where the index corresponding to the class is 1, and the others zero.

> **This simple network performs very well**
> We note that this simple network can without further enhancement exceed the early published results for the MNIST challenge, using standard un-augmented neural networks, simply by using a very wide hidden-layer, that was not feasible at the time of those benchmarks. It is still outperformed by convolution neural networks and other better architectures.

To go into details about each layer: The input is given by the grey scale pixel intensities in the original 28×28 image. This is flattened giving a 784 element sized vector ($\tilde{x}$). The first weight matrix projects that up onto a hidden layer of width 1024. Thus W^1 is a 1024×784 matrix. The bias $\tilde{b}^1$ is for that hidden layer, so is a vector of length 1024. The hidden layer's actual activation values ($\tilde{z}$) for any input can be considered as being the values taken by 1024 different latent variables describing that input. These are chosen from a continuous abstract space of variables (via training) to effectively discriminate the correct class in the next layer (the output layer).

This is a shallow network, with only one hidden layer. A deep network would have more hidden layers, for additional latent variables describing relations. For the output layer we must project down to 10 dimensions, as there are 10 classes to choose from: the numbers from 0 to 9. thus W^2 is a 10×1024 matrix, and the bias $\tilde{b}^2$ is a 10 element vector. Using the softmax layer here ensures that output $\hat{y}$ is a valid probability mass function (nonnegative and summing to one).

1.5.1.1   Softmax and Bayes' Theorem

As a digression, it is worth considering the similarity between a network with a softmax output layer, and the application of Bayes' Theorem. This will become important for understanding output embeddings, and hierarchical softmax in the future chapters.

The conditional probability of a classification given the value of some observed variable is defined by $P(Y = y \mid Z = \tilde{z})$. For Y being the class taking value i; and Z the variable being conditioned upon, taking value $\tilde{z}$. Here $\tilde{z}$ is some feature vector, this could be the output of a hidden layer below, or it could be a direct input. Using softmax the estimated probability is given by

$$P(Y = i \mid Z = \tilde{z}) = \text{smax}(\tilde{z})_i \qquad (1.18)$$

$$= \frac{\exp\left(W\tilde{z} + \tilde{b}\right)_i}{\sum_{\forall j} \exp\left(W\tilde{z} + \tilde{b}\right)_j} \qquad (1.19)$$

$$= \frac{\exp{(W\tilde{z})}_i \, \exp\left(\tilde{b}_i\right)}{\sum_{\forall j} \exp{(W\tilde{z})}_j \, \exp\left(\tilde{b}_j\right)} \qquad (1.20)$$

One can see that the bias term $\exp \tilde{b}_i$ does not depend on the value of z. The bias term is analogous to the prior probability. Literally, it is the bias towards each element (i.e. each value Y could take) without observing the condition. We will define an unnormalised probability-like scoring function R; and say $R(Y = i) = \exp{(\tilde{b}_i)}$, representing the marginal score contribution towards class i from the bias.

The other component is $\exp{(W\tilde{z})}_i$. By considering this for each index i (class value) that Y might take then we have

$$(W\tilde{z})_i = \sum_{\forall j} W_{i,j} \, \tilde{z}_j = W_{i,:}\tilde{z} \qquad (1.21)$$

Given one is considering the case for a particular class $Y = i$, then it can be seen that the row vector $W_{i,:}$ as a weighting map for features possessed by $\tilde{z}$, i.e. for input (Z) feature j, $W_{i,j}$ determines the relative weighting for jth feature versus the other features of Z, when the class is i. When trained, the weight values will be such that it scores how likely the features are to occur for a given output class.

We can say that our scoring function R, has a likelihood term:

$$R(Z = \tilde{z} \mid Y = i) = \exp{(W\tilde{z})}_i \qquad (1.22)$$

We will return to these row vectors $W_{i,:}$ when we discuss as output embeddings in Chap. 3. $W_{i,:}$ must characterise the output class i, since difference classes would assign different importance to different features in the Z. Furthermore, to some degree similar classes, i.e. classes triggered by similar features, would thus have more similar weights.

We can combine the unnormalised score factors to reformulate the original statement:

$$P(Y = i \mid Z = \tilde{z}) = \frac{R(Z = \tilde{z} \mid Y = i) \, R(Y = i)}{\sum_{\forall j} R(Z = \tilde{z} \mid Y = j) \, R(Y = j)} \qquad (1.23)$$

Contrast this to Bayes' Theorem:

$$P(Y = i \mid Z = \tilde{z}) = \frac{P(Z = \tilde{z} \mid Y = i) \, P(Y = i)}{P(Z = \tilde{z})} \qquad (1.24)$$

$$= \frac{P(Z = \tilde{z} \mid Y = i) \, P(Y = i)}{\sum_{\forall j} P(Z = \tilde{z} \mid Y = j) \, P(Y = j)} \qquad (1.25)$$

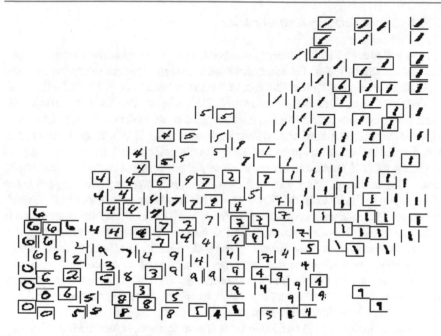

Fig. 1.9 A sampling of MNIST images from the test-set, arranged according to the values of 2 neurons on the encoding layer

---

Marginal Denominator in Bayes Theorem
In Eq. (1.25), notice that we replace $P(Z = \tilde{z})$. We do this using the rule for finding marginal probabilities, from conditional probabilities with mutually exclusive and exhaustive condition values. This is always possible when working with classification.

The similarities can be seen. The bias effectively determines a prior probability-like term. The weights defines the posterior probability: that is the chance of a particular class having a particular set of features. The rows of the weight matrix mark how important each hidden feature is to each class. We can consider that each row of the weight matrix is itself a representation of the class, as characterised by the importance of those latent features. In the next chapter we will consider this as an output embedding (Fig. 1.9).

A more traditional way of finding a representation of an input (rather than an output class) is to use an autoencoder.

1.5.2 Bottlenecking Autoencoder

An autoencoder is a tool primarily used for finding a representation of their input. There are many varieties of autoencoder based on neural network related techniques, including the works of Hinton (2002), Hinton and Salakhutdinov (2006), Hinton et al. (2006), Vincent et al. (2010), Chen et al. (2012), Kingma and Welling (2014). This itself is a whole sub-field of machine learning. Here we look at a simple bottlenecking autoencoder (Bourlard et al. 1988; Japkowicz et al. 2000). It has been used in a variety of tasks to attempt to find an optimal representation for an input e.g. as in Usui et al. (1992). An autoencoder is a neural network tasked with outputting its input. This seemingly is a pointless task – one already has the perfect reproduction of the input, in the input itself. However, the true use of an autoencoder is to extract the output of one of the intermediary layers. We call the intermediary layer the code layer. To force this coded representation to have useful interesting properties, and to prevent the network from simply learning the identity function, all autoencoder include one or more features that increases the difficulty. In the case of the bottlenecking autoencoder this feature is the bottleneck. The code layer is much narrower than the input layer. This forces the network to effectively learn to compress the data – performing dimensionality reduction.

In this particular example we are looking at an auto-encoder for the MNIST images discussed earlier. The original images are 28×28 pixels, i.e. 784 dimensional. We compress it down to just 2 dimensions using the network shown in Fig. 1.10.

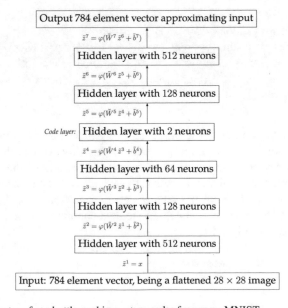

Fig. 1.10 A structure for a bottle necking autoencoder for use on MNIST

Fig. 1.11 The Leaky ReLU6 function. The leak level on this plot is exaggerated for purposes of visualisation

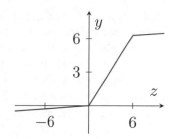

Fig. 1.12 Recreation of an input from the MNIST test set. Input is on the left, output is on the right

In this particular network φ is a leaky RELU6 unit. As shown in Fig. 1.11

$$\varphi(z) = \begin{cases} 0.01z + 6 & 6 < z \\ 1.01z & 0 \le z \le 6 \\ 0.01z & z < 0 \end{cases} \tag{1.26}$$

The reason for this is that a sigmoidal unit does not perform well in a deep network because of the gradient-vanishing/exploding effects. As a result, only the bias information of the final layer mattered, effectively making all outputs just the average of all inputs. The normal solution to this in deep networks is ReLU or ReLU6 units.

However, a ReLU, or ReLU6 unit is not ideal either, as these units turn-off, and cannot turn back on again. If one of the only 2 neurons in the code layer turn-off then they cannot turn back on again – thus forcing it down to just one neuron, or none if that neuron also dies. In trialling this network structure, this was found to occur quiet very often.

The solution was to add a "leak" to the unit. A small constant gradient even in the saturated position. Making such a neuron possible, even if difficult, to turn back on once turned off. With this change the network always produces quality results. An example of the recreation of an input from the test set can be seen in Fig. 1.12. It can be seen that the blur suggests that the input is a 7 or 9 like shape – that level of information survived being squeezed through the bottleneck.

More significantly, a sampling of the code layer is visually shown in Fig. 1.9. In this image, the positions on the X and Y axis for each input image is given by the values from the code layer for the representation of that image. It can be seen that the code is capturing key information about the appearance. The 0s appear near to the similarly rounded 6s, the 8s near the 3s, etc. In particular the 1s are arrayed according to their slope. The autoencoder has captured very useful information about the inputs, that would be hard to define with any hand-written feature extractor.

It is this property of networks, to implicitly discover and define the most important latent variables and relationships to the task at hand that also makes it valuable for natural language processing.

References

Bengio, Yoshua, Pascal Lamblin, Dan Popovici, and Hugo Larochelle. 2007. Greedlayer-wise training of deep networks. In *Advances in neural information processing systems*, 153–160.

Bourlard, Herve, and Yves Kamp. 1988. Auto-association by multilayer perceptronsand singular value decomposition. *Biological Cybernetics* 59 (4): 291–294.

Chen, Minmin, Zhixiang Xu, Kilian Weinberger, and Fei Sha. 2012. Marginalized denoising autoencoders for domain adaptation. In *Proceedings of the 29th international conference on machine learning (ICML-12)*, eds. John Langford and Joelle Pineau, 767–774. Edinburgh: Omnipress. ISBN: 978-1-4503-1285-1.

Dahl, George E., Tara N. Sainath, and Geoffrey E. Hinton. 2013. Improving deep neural networks for LVCSR using rectified linear units and dropout. In *2013 IEEE international conference on acoustics, speech and signal processing (ICASSP)*, 8609–8613. IEEE. https://doi.org/10.1109/ICASSP.2013.6639346.

Dauphin, Yann N., Harm de Vries, Junyoung Chung, and Yoshua Bengio. 2015. RMSProp and equilibrated adaptive learning rates for non-convex optimization. In *CoRR*. arXiv:abs/1502.04390.

Duchi, John, Elad Hazan, and Yoram Singer. 2011. Adaptive subgradient methodsfor online learning and stochastic optimization. *The Journal of Machine Learning Research* 12: 2121–2159.

Glorot, Xavier, Antoine Bordes, and Yoshua Bengio. 2011. Deep sparse rectifier neural networks. In *Proceedings of the fourteenth international conference on artificial intelligence and statistics*, 315–323.

Goodfellow, Ian, Yoshua Bengio, and Aaron Courville. 2016. *Deep learning*. MIT Press. http://www.deeplearningbook.org

Hahnloser, Richard L.T. 1998. On the piecewise analysis of networks of linearthreshold neurons. *Neural Networks* 11 (4): 691–697.

Hebb, D.O. 1949. *The organization of behavior: A neuropsychological theory*. A Wiley book in clinical psychology. New York: Wiley.

Hinton, Geoffrey E. 2002. Training products of experts by minimizing contrastive divergence. *Neural Computation* 14 (8): 1771–1800.

Hinton, Geoffrey E., Simon Osindero, and Yee-Whye Teh. 2006. A fast learning algorithm for deep belief nets. *Neural Computation* 18 (7): 1527–1554

Hinton, Geoffrey E. and Ruslan R. Salakhutdinov. 2006. Reducing the dimensionality of data with neural networks. *Science 313* (5786): 504–507.

Japkowicz, Nathalie, Stephen Jose Hanson, and Mark A. Gluck. 2000. Nonlinear auto association is not equivalent to PCA. *Neural Computation* 12 (3): 531–545.

Kingma, D. P. and M. Welling. 2014. Auto-encoding variational bayes. In *The international conference on learning representations (ICLR)*. arXiv:1312.6114 [stat.ML].

Kingma, Diederik and Jimmy Ba. 2014. Adam: A method for stochastic optimization. arXiv:1412.6980.

Krizhevsky, Alex and G. Hinton. 2010. Convolutional deep belief networks on CIFAR-10. *Unpublished Manuscript 40*.

LeCun, Yann A., Leon Bottou, Genevieve B. Orr, and Klaus-Robert Muller. 2012. Efficient backprop. *Neural networks: Tricks of the trade*, 9–48. Berlin: Springer.

Leshno, Moshe, Vladimir Ya. Lin, Allan Pinkus, and Shimon Schocken. 1993. Multilayer feedforward networks with a nonpolynomial activation function can approximate any function. *Neural Networks* 6 (6): 861–867.

Mhaskar, Hrushikesh N., and Charles A. Micchelli. 1992. Approximation by superposition of sigmoidal and radial basis functions. *Advances in Applied Mathematics* 13 (3): 350–373.

Minsky, Marvin, Seymour A. Papert, and Leon Bottou. 2017. *Perceptrons: An introduction to computational geometry*. Cambridge: MIT Press.

Ngiam, Jiquan, Adam Coates, Ahbik Lahiri, Bobby Prochnow, Quoc V. Le, and Andrew Y. Ng. 2011. On optimization methods for deep learning. In *Proceedings of the 28th international conference on machine learning (ICML-11)*, 265–272.

Nielsen, Michael A. 2015. *Neural networks and deep learning*. Determination Press.

Nocedal, Jorge. 1980. Updating quasi-Newton matrices with limited storage. *Mathematics of Computation* 35 (151): 773–782.

Rumelhart, David E., Geoffrey E. Hintont, and Ronald J. Williams. 1986. Learning representations by back-propagating errors. Nature 323: 9.

Sonoda, Sho and Noboru Murata. 2017. Neural network with unbounded activation functions is universal approximator. *Applied and Computational Harmonic Analysis 43* (2), 233–268. ISSN: 1063-5203, https://doi.org/10.1016/j.acha.2015.12.005.

Usui, Shiro, Shigeki Nakauchi, and Masae Nakano. 1992. Reconstruction of Munsell color space by a five-layer neural network. *Journal of the Optical Society of America A* 9 (4): 516–520. https://doi.org/10.1364/JOSAA.9.000516.

Vincent, Pascal, Hugo Larochelle, Isabelle Lajoie, Yoshua Bengio, and Pierre- Antoine Manzagol. 2010. Stacked denoising autoencoders: Learning useful representations in a deep network with a local denoising criterion. *Journal of Machine Learning Research* 11: 3371–3408.

White, Lyndon. 2016. Encoding angle data for neural networks. Cross validated stack exchange. https://math.stackexchange.com/q/2369786.

Xiao, Han, Kashif Rasul, and Roland Vollgraf. 2017. Fashion-MNIST: A novel image dataset for benchmarking machine learning algorithms. arXiv:1708.07747 [cs.LG], https://github.com/zalandoresearch/fashionmnist.

Zeiler, Matthew D. 2012. ADADELTA: An adaptive learning rate method. In *CoRR*. arXiv:abs/1212.5701.

Recurrent Neural Networks for Sequential Processing

2

> *I've the RNN with and works, but the computed*
> *with program of the RNN with and the*
> *computed of the RNN with with and the code —*
> The output of an RNN created by Andrej
> Karpathy (2015), trained on an article on the
> use of RNNs for generating text (it works poorly
> due to the very low amount of training data)
> http://karpathy.github.io/2015/05/21/rnneffectiveness/

Abstract

This chapter continues from the general introduction to neural networks, to a focus on recurrent networks. The recurrent neural network is the most popular neural network approach for working with sequences of dynamic size. As with the prior chapter, readers familiar with RNNs can reasonably skip this. Note that this chapter does not pertain specifically to NLP. However, as NLP tasks are almost always sequential in nature, RNNs are fundamental to many NLP systems

Works generated by RNNs
Generating text using RNNs is very hip, and pretty rad. https://arstechnica.com/?p=896589, https://github.com/zackthoutt/got-book-6, http://aiweirdness.com/post/168770625987/. The results are often surprisingly cognisant, and almost always entertaining. It should be noted however, that this is in-principle little different to sampling Markov chains from probabilistic language models. Even the longest of longer short term memory systems still do not truly have the memory to actually sensibly write prose; which would require memory of the outputs from mutliple paragraphs (or chapters) ago.

© Springer Nature Singapore Pte Ltd. 2019
L. White et al., *Neural Representations of Natural Language*,
Studies in Computational Intelligence 783,
https://doi.org/10.1007/978-981-13-0062-2_2

2.1   Recurrent Neural Networks

Time-Step

RNNs are normally described in terms of a time-step. This is the advancement of the system such that the previous output state, is now the input old-state. This does not have to be literal time – indeed it cannot be, as it is discrete. In most NLP applications it is time analogous: words in the order they are said. In other NLP applications it might actually be words in the reverse order to that in which they are normally said. In other machine learning applications it may not correspond to time at all. For example, using a rotating distance sensor (e.g radar) each time-step corresponds to a different angle of the antenna.

A key limitation of a neural network is that the number of inputs and outputs must be known at training time and must always be the same for all cases. This is not true for natural languages: if a problem involves processing a sentence, then each input will be made up of a varying number of words. Similarly for the output, in a text generation case.

Recurrent neural networks (RNN) overcome this by allowing the network to have a state that persists over time. Inputs of any size can be handled one fixed-sized part of the input (e.g. one word, or one frame from a video) at a time, using the state to remember the past inputs.

A RNN is effectively a chain of feed-forward neural networks, each one being identical in terms of their weight and bias parameters. While identical in terms of parameters, they each act at a different time-step.

At each time-step the same network is used, with different inputs. For purposes of looking at this in the big picture, we will first consider each network as a black-box recurrent unit (RU). The recurrent unit takes at each time-step, an input for that time-step, some representation of the state for the RU at the previous time-step; and produces an output for this time-step, and the state representation to be used in the next time-step. A diagram of this is shown in Fig. 2.2. It is worth distinguishing that the unit output is not the same as the overall network output, it is just the output of this sub-network at this time-step. Each time-step can be considered as having two effective inputs (the previous state, and the actual input) and two effective outputs (next state and actual output).

Not all of these inputs and outputs are actually used at all time-steps meaningfully. The initial state for the first time-step is normally set to some zero vector, and the final state at the end of the sequence is normally discarded.

2.2   General RNN Structures

In general most popular uses of recurrent networks belong to one of four general types of structure. Matched-sequence, encoder, decoder, and encoder-decoder. These common structures are shown in Fig. 2.1. The motivation for using an RNN comes from needing to solve problems where the size of the input and/or the size of the output is not consistent across all cases.

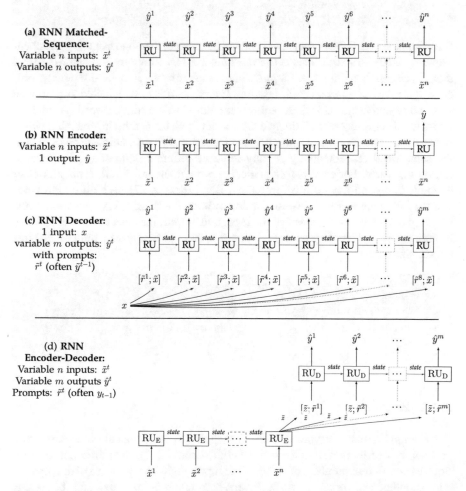

Fig. 2.1   The unrolled structure of an RNN for **a** Matched-sequence **b** Encoding, **c** Decoding and **d** Encoding-Decoding (sequence-to-sequence) problems. RU is the recurrent unit – the neural network which reoccurs at each time-step

2.2.1 Matched-Sequence

Memory
The term memory refers to the state. A RNN can be said to remembers something if the information about that is encoded into its state. One should be clear on the difference between learning/training (done by all NNs over the whole the training data) and remembering (done by RNNs within the processing of a single item, across multiple time-steps, during execution).

If the input size and the output size is always the same, then one can use a matched-sequence RNN structure. This is the most basic RNN structure. At each time-step, there is an input, and a target output. An example of this, in natural language processing, is Part of Speech (POS) tagging. Every word is to be labelled as a noun, a verb, an adjective etc. This does require "memory" as the context words around the target word being classified influence the correct POS. Remembering the other words (from other time-steps) is required to use the context to disambiguate cases where the same word can potentially occupy multiple different parts of speech depending on the usage. For example record is both a noun and a verb. This particular example is a good use for a Bidirectional RNN (Sect. 2.4.2), as both the previous and the following words are useful for determining the POS. A key limitation of the matched-sequence structure is that the input and output size must be the same. Often though one would like to process an input of that could be any size, but produce just a fixed size output.

Overlapping Parts of Speech
It is nearly ubiquitous that verbs and adjectives have a noun form. Words that only occupy a single part of speech are the exception rather than the norm, in English.

2.2.2 Encoder

For example if one is trying to learn a mapping from a textual color name to a probability distribution in color-space (White et al. 2017), then different descriptions have different numbers of words. One input might be very light green, while another just orange. At each time-step one input is provided – being one of the words in the (potentially multi-word) color name. Another example use is in sentiment analysis, predicting the sentiment being expressed by a text as positive or negative. All the outputs at all time-steps, except the last, can be ignored. The output of the final time-step can be connected to a further network with a final output layer giving the overall output. The situation, with a variable number of inputs, but a fixed size of output is described as an encoder network.

2.2.3 Decoder

Pseudo-tokens

Pseudo-tokens such as <EOS> (end of string), <START> (start of string), and similar are common in RNN tasks. As mentioned, outputting <EOS> is required to be known when the decoder is done outputting. <START> can be useful to model things that occur near the start of input; and as an initial prompt. In some tasks other non-word tokens might be inserted also, such as in transcription of recorded speech, a <PAUSE> token might be included. Pragmatically, as long as the symbols used to represent these never occur amongst the true word tokens, these can be treated just like regular words by the system.

The reverse is a decoder network. This means a fixed sized input being mapped to a variably sized output. If the system is attempting to learn from a point in the color space to the name of that color (Monroe et al. 2016). For example: (144,238,144) very light green (the outputs); but (255,165,0) might map to just orange (one output). In this decoder type network, there is one true input, at the first time-step, and the output from every time-step is used. Each output can be connected to a softmax layer giving a probability for possible words. Even though the decoder only has a fixed sized input, never-the-less an input must still be provided at every time-step. We call this input a prompt, as it is not providing new information to the network, merely driving it to produce the next output. It is common to include an end of string marker token (often literally <EOS>), so as to know when to stop prompting for additional outputs.

2.2.3.1 Prompts in Decoder RNNs

Terminology: Prompt

The word prompt is our own terminology. We are not aware of a consistent term used in the literature for the input at each time-step to a decoder RNN. The phrase "dummy input" would also work, although as discussed the choice of prompt can allow useful information to be added. Simplifying the learning problem.

Prompts are required for decoders as the network must have some input to cause it to give an output. In theory this could be a constant: all inputs and all outputs for those inputs, can be learned and remembered by the networks memory (encoded in its state). In practice it is very common to include part of the output of the previous step as part of the prompt. When generating a sequence of words for example, one can use the previous word. At training time this can be the targeted previous output (y^{i-1}). At test time (and in real deployment) this is normally the most-likely predicted output (argmax $\hat{y}^{t-1}$). This effectively gives direct bigram state information to the model, allowing the memory to focus on higher level tasks. It also allows the outputs to be explored, for example by providing the second most-likely word as the first prompt, a different sequence can be generated.

It is also common to include in prompts the original input to the decoder. For example in a caption generator, including a vector representation of the image (for example an Inception Image Embedding Szegedy et al. 2015); in the color decoder example this would be including the original color representation. In general the prompt can be used to add information to the network ensuring that each time-step can do as well as possible, even if the state does not capture all the desired information. Ensuring the state is able to capture all information is part of designing the internals of the RU.

2.2.4 Encoder-Decoder

The encoder-decoder RNN is the generalised structure for sequence to sequence learning (Cho et al. 2014), it is sometimes called a seq2seq model. It should be contrasted to the matched-sequence RNNs takes a sequence as an input and produces a sequence *of the same size* as an output. The restriction to the sequences being the size limits to the of the matched-sequence RNN for many tasks. For example, in an application such as machine translation, question answering, or video captioning, a sequential input and output is required, but the sequences have different lengths. A sentence in one language will not normally translate to a sentence with the exact same number of word in another. The solution to this is the encoder-decoder RNN. In this structure an encoder RNN is used to take the input, and its final output is then connected as the input to a separate decoder RNN. Thus separating the input processing from the output generation.

> **Output/Input layers**
> Extra feed-forward layers are often applied between the network's overall input and output, and the RU's input an output. The extra layers could be considered as part of the RU, as they occur at each time-step (even if not used). Alternatively they can be considered as something that surrounds the RU, only at certain time-steps. This depends on the point of view. For our purposes we will consider the core of the RU in isolation.

2.3 Inside the Recurrent Unit

In this section we discuss the various different types of Recurrent Unit (RU). The type of RU determines the different types of recurrent networks, such as Elman

Fig. 2.2 A recurrent unit
with its 2 inputs and 2
outputs. Not shown are the
internal functioning which
may be a complex (e.g. as in
LSTM) or simple (e.g. as in
a Basic RU) neural network
of its own

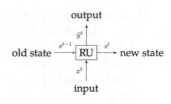

networks, GRU networks and LSTM networks. As discussed, and shown in Fig. 2.2,
every recurrent unit from the outside has the previous state, the next state, the unit
input and the unit output. What differs is how they are connected and what controls
the information within them. Every recurrent unit is itself a neural network.

In many types of RU (e.g. GRU, Basic RU) the output and the state are always
equal. This particularly makes sense when they should be capturing the same kinds
of information (as in a decoder-encoder). Furthermore, as there will be additional
feed-forward layers on top of used outputs (if nothing else an output layer is normally
required), the need to differentiate output from state is lessened. However, it is a distinction made in the very well known LSTM unit (Sect. 2.3.3) so we preserve it here.

2.3.1   Basic Recurrent Unit

The most basic recurrent unit, is a single layer $\tilde{h}^t$, with the unit output, and the unit
state both being the value of this layer. The layer is not hidden from the perspective
of the recurrent unit, but is from the perspective of the whole network. The network
inside the basic recurrent unit, shown in Fig. 2.3, is given by:

$$\tilde{h}^t = \varphi \left(W \left[\tilde{x}^t ; \tilde{s}^{t-1} \right] + \tilde{b} \right) \tag{2.1}$$

$$\tilde{s}^t = \tilde{h}^t \tag{2.2}$$

$$\hat{y}^t = \tilde{h}^t \tag{2.3}$$

where W and $\tilde{b}$ are the weight matrix and bias vector obtainable from training.

Fig. 2.3 A basic recurrent
unit

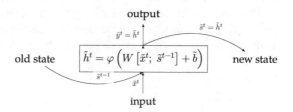

One could call this an Elman RU: the networks considered in Elman and Jeffrey (1990) are such a basic RU, with an extra overall output layer on top.

Basic RUs are not used in many modern works. It is difficult for these networks to propagate error information across many time-steps (Bengio et al. 1994). This results in the state not capturing information from many time-steps ago. Thus the network has a very short memory.

Other more advanced recurrent units solve this problem by placing more explicit controls on the state.

Jordan RU
The Jordan Network formulation is not commonly used today, however analogously to considering the Basic RU as an Elman RU, a similar formulation can be done for a Jordan network (Jordan and MI 1986). The Jordan RU would be given by:

$$\tilde{h}^t = \varphi \left(W \left[\tilde{x}^t; \; \tilde{s}^{t-1} \right] + \tilde{b} \right)$$
$$\tilde{o}^t = \varphi \left(V \tilde{h}^t + \tilde{c} \right)$$
$$\tilde{s}^t = \tilde{o}^t$$
$$\hat{y}^t = \tilde{o}^t$$

One of these RUs alone is a Jordan network (it does not need an additional output layer). As an RU the difference is in the additional layer $\tilde{o}^t$ without direct access to the previous state.

2.3.2 Gated Recurrent Unit

The Gated Recurrent Unit (GRU) was introduced by Cho et al. (2014a). The GRU is actually a simplification of the much older and better known Long Short Term Memory (LSTM), which will be discussed in Sect. 2.3.3. We discuss the GRU first as it is the simpler system. In the evaluations of Chung et al. (2014) and Jozefowicz et al. (2015) it was found to perform very well – similar to LSTM for most tasks.

The core insight is to gate the changes in the state. There are subnetworks within the RU that we call gates. These subnetworks learn how the state should change. Thus helping the overall network preserve better information in the state.

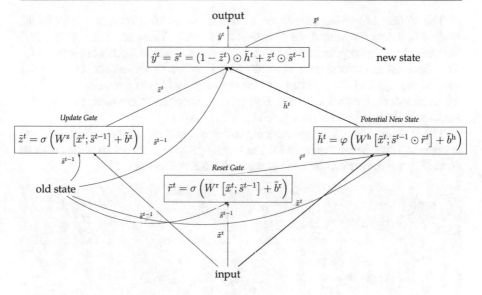

Fig. 2.4 A gated recurrent unit

The gated recurrent unit (GRU), as shown in Fig. 2.4, is defined by:

$$\tilde{z}^t = \sigma \left(W^z \left[\tilde{x}^t; \tilde{s}^{t-1} \right] + \tilde{b}^z \right) \tag{2.4}$$

$$\tilde{r}^t = \sigma \left(W^r \left[\tilde{x}^t; \tilde{s}^{t-1} \right] + \tilde{b}^r \right) \tag{2.5}$$

$$\tilde{h}^t = \varphi \left(W^h \left[\tilde{x}^t; \tilde{s}^{t-1} \odot \tilde{r}^t \right] + \tilde{b}^h \right) \tag{2.6}$$

$$\tilde{s}^t = (1 - \tilde{z}^t) \odot \tilde{h}^t + \tilde{z}^t \odot \tilde{s}^{t-1} \tag{2.7}$$

$$\hat{y}^t = \tilde{s}^t \tag{2.8}$$

where W^z, W^r and W^h are the weight-vectors, and $\tilde{b}^z$, $\tilde{b}^r$ and $\tilde{b}^h$ are the bias vectors for the 3 separate layers.

Trade-off formula

It may be stating the obvious, but a standard way (and the only sensible way) to trade-off a value between two possible choices: a and b, when given a preference level for the first as p (a number between zero and one) is $(p)a + (1 - p)b$. If p is interpreted as the probability of a being the correct value of a random variable, with b as the other possible value, then this is the expected value of that random variable.

This may look complex, but it can be broken down into parts. First we can see that, like in the basic RU, the output and the state have the same value (Eq. 2.8).

The $\tilde{h}^t$ (Eq. 2.6) is our core layer, as in the basic cell. However, unlike in the basic cell, it does not immediately become the state: $\tilde{s}^t$. There is a trade-off (Eq. 2.7) between the state keeping its old value $\tilde{s}^{t-1}$, and getting the hidden layer value $\tilde{h}_t$. This trade-off is element-wise controlled by $\tilde{z}^t$ which is valued between 0 and 1 (due to the sigmoid unit). When an element of $\tilde{z}^t$ is 1, then the value of $\tilde{s}^{t-1}$ for the corresponding element is kept. Conversely, when the $\tilde{z}^t$ element is 0, then the element of the new state is fully given by $\tilde{h}^t$.

The $\tilde{z}^t$ is often called the *update gate* as it controls ("gates") how much the state is updated using the new value in $\tilde{h}^t$. The update-gate sub-network uses the previous state $\tilde{s}^{t-1}$ and the present input $\tilde{x}^t$ to make this determination.

Multiplication, Concatenation, and Addition

It is important to grasp that the product of a matrix and the concentration of two vectors can also be expressed as the sum of the product of a block of that matrix. They are the same thing. $W \cdot \left[\tilde{a}; \tilde{b}\right] = U\tilde{a} + V\tilde{b}$ if $W = [U \; V]$ The difference is purely notation.

The *reset gate* is loosely similar to the update gate. $\tilde{r}^t$ controls how much influence the past state $\tilde{s}^{t-1}$ has on calculating the new value of $\tilde{h}^t$ – which is the new potential state/output as discussed. It is perhaps clearer if $\tilde{h}^t$ is reformulated to split V into the terms which multiply with $\tilde{x}^t$ and the terms which multiply with $\tilde{s}^{t-1}$.

For $W^h = \left[W^{hx} \; W^{hs}\right]$ we can write:

$$\tilde{h}^t = \varphi \left(W^h \left[\tilde{x}^t; \tilde{s}^{t-1} \odot \tilde{r}^t\right] + \tilde{b}^h \right) \tag{2.9}$$

$$= \varphi \left(W^{hx}\tilde{x}^t + W^{hs}\left(\tilde{s}^{t-1} \odot \tilde{r}^t\right) + \tilde{b}^h \right) \tag{2.10}$$

The $\tilde{r}^t$ is called the reset-gate because it wipes the effect of the old-state in calculating the potential new state $\tilde{h}^t$. When $\tilde{r}^t$ is reduced to zero, then the updated value for $\tilde{h}^t$ is as if $\tilde{s}^{t-1}$ was just like for the initial zero-vector state – it is zeroed out by $\tilde{r}^t$. If the update gate $\tilde{z}^t$ is high then it would fully reset the system.

Notice that when $\tilde{z}^t = 0$ and $\tilde{r}^t = 1$, then the system is identical to the Basic RU. This could be achieved by setting suitably large biases. However, the system is more flexible than that, since $\tilde{z}^t$ and $\tilde{r}^t$ are themselves very similar to Basic RUs – though they do not control their own state. The gates can behave differently based on the inputs and states to recognise the most important information that must be stored.

2.3.3 LSTM Recurrent Unit

The LSTM is the most well known RNN unit. The term is very nearly interchangeable with RNN today. The original form was proposed by Hochreiter et al. (1997). The form in current use is a variant from Gers et al. (1999).

LSTM uses a compound state, comprised of the units previous output, and an additional state vector called the *cell*. We write $\tilde{s}^t = (\tilde{c}^t, \hat{y}^t)$, to fit the normal formulation, though for maths it is easier to work with these in parts.

As shown in Fig. 2.5, the LSTM RU is defined by:

$$\tilde{s}^t = (\tilde{c}^t, \hat{y}^t) \tag{2.11}$$

$$\tilde{i}^t = \sigma \left(W^i \left[\tilde{x}^t; \hat{y}_{t-1} \right] + \tilde{b}^i \right) \tag{2.12}$$

$$\tilde{f}^t = \sigma \left(W^f \left[\tilde{x}^t; \hat{y}_{t-1} \right] + \tilde{b}^f \right) \tag{2.13}$$

$$\tilde{o}^t = \sigma \left(W^o \left[\tilde{x}^t; \hat{y}_{t-1} \right] + \tilde{b}^o \right) \tag{2.14}$$

$$\tilde{h}^t = \tanh \left(W^h \left[\tilde{x}^t; \hat{y}_{t-1} \right] + \tilde{b}^h \right) \tag{2.15}$$

$$\tilde{c}^t = \tilde{i}^t \odot \tilde{h}^t + \tilde{f}^t \odot \tilde{c}^{t-1} \tag{2.16}$$

$$\hat{y}^t = \tilde{o}^t \odot \varphi(\tilde{c}^t) \tag{2.17}$$

In LSTM there are three gate sub-networks: the input gate $\tilde{i}^t$, the forget gate $\tilde{f}^t$, and the output gate $\tilde{o}^t$.

Together the input and forget-gates take the purpose of the GRU's update-gate. Consider the case if $\tilde{o}^t = 1$ and φ is the identify function, and with $\tilde{i}^t = 1 - \tilde{f}^t$, that would make the value for $\tilde{c}^t$ very similar to GRU's $\tilde{s}^t$ (and $\hat{y}^t$ identical).

Individually, the forget gate $\tilde{f}^t$ controls the extent that the previous value of the cell is used, and the input gate $\tilde{i}^t$ controls the extent to which the new potential value $\tilde{h}^t$ is used for the new value of $\tilde{c}^t$.

The output-gate's obvious purpose is to gate the output y_t. However, as the output forms part of the state, this has an effect on the networks next time-step. Loosely this can be seen as similar to the GRU's reset gate. If we substitute the value for $\hat{y}_{t-1} = \tilde{o}^{t-1} \odot \varphi(\tilde{c}^{t-1})$ into $\tilde{h}^t$:

$$\tilde{h}^t = \tanh \left(W^h \left[\tilde{x}^t; \hat{y}_{t-1} \right] + \tilde{b}^h \right) \tag{2.18}$$

$$\tilde{h}^t = \tanh \left(W^h \left[\tilde{x}^t; \tilde{o}^{t-1} \odot \varphi(\tilde{c}^{t-1}) \right] + \tilde{b}^h \right) \tag{2.19}$$

it can be seen that, at the previous time-step, setting an element of o_{t-1} to zero effectively removes the effect of the previous cell-state from the equations for all the gates and the the potential new cell $\tilde{h}^t$. Thus resetting the network. (Similar substitutions can be done for $\tilde{i}^t$, $\tilde{f}^t$ and $\tilde{o}^t$)

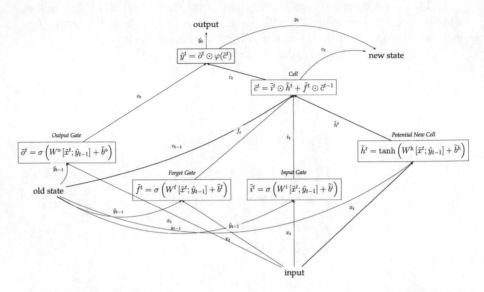

Fig. 2.5 A LSTM recurrent unit

Biasing the Forget Gate
Practically, it is important to initially set the forget gate bias to 1. This is unlike all the other initialisation in the networks to 0 (or a small random number e.g. `randn(0.01)`). Equivalently one can also just add a constant additional bias term of $+1$ to the forget gate equations. This has been show to benefit almost all uses of LSTM (Gers et al. 1999; Jozefowicz et al. 2015). This is the default in most neural network libraries (though only recently patched in some).

2.4   Further Variants

2.4.1   Deep Variants

One can have a deep RNN. This is done by stacking additional recurrent units above the existing ones. Attaching, the unit output as the unit input of the layer above. The result can be interpreted as a single, deep recurrent unit.

2.4.2   Bidirectional RNNs

A Bidirectional RNN is similar to deep variants (Schuster et al. 1997). This is effectively using two RNNs so that the input can be processed from both temporal directions. The forwards and backwards RNNs both take the same input at each time-step, but are otherwise fully distinct. The unit outputs at each time-step are concatenated to give an overall time-step output from the RUs.

This is not so useful in most decoder RNNs, the number of outputs (and thus time-steps) of a decoder RNN is not known except by running them until an end-marker is output. However, there are absolutely no issues when using them for an encoder. In an encoder network the whole input is available before the output needs to be produced. Similarly for a matched-sequence RNN, if the sequence can be broken down into blocks (e.g. sentences), and a real-time output is not required.

2.4.3   Other RNNs

There also exist many other RNNs. Some are similar in structure to those discussed and can be analysed similarly via their recurrent units. Such as the two models of Mikolov et al. (2014), which incorporate a decaying sum of past states into the current state (either gated, or constant). Other networks incorporate various other data-structures (or analogies to such structures) into their recurrent structure: such as stacks (Dyer et al. 2015), pointers/arrays (Weston et al. 2014), or tapes (Graves et al. 2014).

As natural language is primarily a sequential task, RNNs in various forms will be discussed in almost every chapter of this book.

References

Bengio, Yoshua, Patrice Simard, and Paolo Frasconi. 1994. Learning long-term dependencies with gradient descent is difficult. *IEEE Transactions on Neural Networks* 5 (2): 157–166.
Cho, Kyunghyun, Bart van Merriënboer, Dzmitry Bahdanau, and Yoshua Bengio. 2014a. On the properties of neural machine translation: Encoder-decoder approaches. In *Eighth workshop on syntax, semantics and structure in statistical translation* (*SSST-8*).
Cho, Kyunghyun, Bart van Merrienboer, Caglar Gulcehre, Dzmitry Bahdanau, Fethi Bougares, Holger Schwenk, and Yoshua Bengio. 2014b. Learning phrase representations using RNN encoder-decoder for statistical machine translation. In *Proceedings of the 2014 conference on empirical methods in natural language processing* (*EMNLP*). Doha, Qatar: Association for Computational Linguistics, 1724–1734.
Chung, Junyoung, Caglar Gulcehre, Kyung Hyun Cho, and Yoshua Bengio. 2014. Empirical evaluation of gated recurrent neural networks on sequence modeling. arXiv:1412.3555.
Dyer, Chris, Miguel Ballesteros, Wang Ling, Austin Matthews, and Noah A. Smith. 2015. Transition-based dependency parsing with stack long short-term memory. In *CoRR*. arXiv:1505.08075.
Elman, Jeffrey L. 1990. Finding structure in time. *Cognitive Science* 14 (2): 179–211.
Gers, Felix A, Jürgen Schmidhuber, and Fred Cummins. 1999. Learning to forget: Continual prediction with LSTM. *Neural Computation* 12 (10): 2451–2471

Graves, Alex, Greg Wayne, and Ivo Danihelka. 2014. Neural turing machines. In *CoRR*. arXiv:1410.5401.

Hochreiter, Sepp, and Jürgen Schmidhuber. 1997. Long short-term memory. *Neural Computation* 9 (8): 1735–1780.

Jordan, M.I. 1986. Serial order: A parallel distributed processing approach (Technical Report No. 8604). San Diego, La Jolla, CA: Institute for Cognitive Science, University of California.

Jozefowicz, Rafal, Wojciech Zaremba, and Ilya Sutskever. 2015. An empirical exploration of recurrent network architectures. In *Proceedings of the 32nd international conference on machine learning (ICML-15)*, 2342–2350.

Mikolov, Tomas, Armand Joulin, Sumit Chopra, Michaël Mathieu, and Marc'Aurelio Ranzato. 2014. Learning longer memory in recurrent neural networks. In *CoRR*. arXiv:1412.7753.

Monroe, W., N. D. Goodman, and C. Potts. 2016. Learning to generate compositional color descriptions. arXiv:1606.03821 [cs.CL].

Schuster, Mike, and Kuldip K. Paliwal. 1997. Bidirectional recurrent neural networks. *IEEE Transactions on Signal Processing* 45 (11): 2673–2681.

Szegedy, Christian, Wei Liu, Yangqing Jia, Pierre Sermanet, Scott Reed, Dragomir Anguelov, Dumitru Erhan, Vincent Vanhoucke, and Andrew Rabinovich. 2015. Going deeper with convolutions. In *The IEEE conference on computer vision and pattern recognition (CVPR)*.

Weston, Jason, Sumit Chopra, and Antoine Bordes. 2014. Memory networks. In *CoRR*. arXiv:1410.3916.

White, L., R. Togneri, W. Liu, and M. Bennamoun. 2017. Learning distributions of meant color. arXiv:1709.09360 [cs.CL].

Word Representations

3

You shall know a word by the company it keeps

J.R. Firth 1957

Abstract

Word embeddings are the core innovation that has brought machine learning to the forefront of natural language processing. This chapter discusses how one can create a numerical vector that captures the salient features (e.g. semantic meaning) of a word. Discussion begins with the classic language modelling problem. By solving this, using a neural network-based approach, word-embeddings are created. Techniques such as CBOW and skip-gram models (word2vec), and more recent advances in relating this to common linear algebraic reductions on co-locations as discussed. The chapter also includes a detailed discussion of the often confusing hierarchical softmax, and negative sampling techniques. It concludes with a brief look at some other applications and related techniques.

The epigraph at the beginning of this section is over-used. However, it is obligatory to include it in a work such as this, as it so perfectly sums up why representations useful for language modelling are representations that capture semantics (as well as syntax).

We begin the consideration of the representation of words using neural networks with the work on language modeling. This is not the only place one could begin the consideration: the information retrieval models, such as LSI (Dumais et al. 1988) and LDA (Blei et al. 2003), based on word co-location with documents would be the other obvious starting point. However, these models are closer to the end point, than they are to the beginning, both chronologically, and in this chapter's layout. From the language modeling work, comes the contextual (or acausal) language model works such as skip-gram, which in turn lead to the post-neural network co-occurrence based works. These co-occurrence works are more similar to the information retrieval

© Springer Nature Singapore Pte Ltd. 2019
L. White et al., *Neural Representations of Natural Language*,
Studies in Computational Intelligence 783,

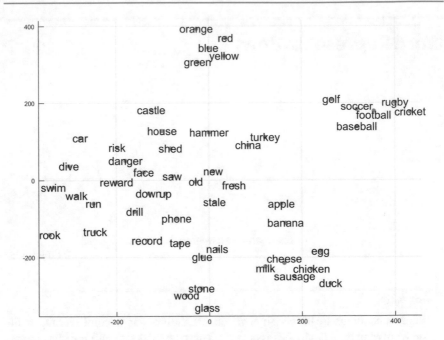

Fig. 3.1 Some word embeddings from the FastText project (Bojanowski et al. 2016). They were originally 300 dimensions but have been reduced to 2 using t-SNE (Maaten and Geoffrey 2008) algorithm. The colors are from 5 manually annotated categories done before this visualisation was produced: foods, sports, colors, tools, other objects, other. Note that many of these words have multiple meanings (see in this chapter), and could fit into multiple categories. Also notice that the information captioned by the unsupervised word embeddings is far finer grained than the manual categorisation. Notice, for example, the separation of ball-sports, from words like run and walk. Not also that china and turkey are together; this no doubt represents that they are both also countries

co-location based methods than the probabilistic language modeling methods for word embeddings from which we begin this discussion.

Word Vector or Word Embedding?
Some literature uses the term *word vector*, or *vector-space model* to refer to representations from LDA and LSA etc. Other works use the terms are used synonymously with *word embedding*. Word embeddings are vectors, in any case.

Word embeddings are vector representations of words. An dimensionality reduced scatter plot example of some word embeddings is shown in Fig. 3.1.

3.1 Representations for Language Modeling

Probability writing convention
We follow convention that capitalised W^i is a random variable, and w^i is a particular value which W^i may take. The probability of it taking that value would normally be written $P(W^i=w^i)$. We simply write $P(w^i)$ to mean the same thing. This is a common abridged (abuse-of) notation. The random variable in question is implicitly given by the name of its value.

The language modeling task is to predict the next word given the prior words (Rosenfeld 2000). For example, if a sentence begins For lunch I will have a hot, then there is a high probability that the next word will be dog or meal, and lower probabilities of words such as day or are. Mathematically it is formulated as:

$$P(W^i=w^i \mid W^{i-1}=w^{i-1}, \ldots, W^1=w^1) \tag{3.1}$$

or to use the compact notation

$$P(w^i \mid w^{i-1}, \ldots, w^1) \tag{3.2}$$

where W^i is a random variable for the ith word, and w^i is a value (a word) it could, (or does) take. For example:

$$P(\text{dog} \mid \text{hot}, \text{a}, \text{want}, \text{I}, \text{lunch}, \text{For})$$

The task is to find the probabilities for the various words that w^i could represent.

Google n-gram corpora
Google has created a very large scale corpora of 1, 2, 3, 4, and 5-grams from over 10^{12} words from the Google Books project. It has been made freely available at https://books.google.com/ngrams/datasets (Lin et al. 2012). Large scale n-gram corpora are also used outside of statistical language modeling by corpus linguists investigating the use of language.

The classical approach is trigram statistical language modeling. In this, the number of occurrences of word triples in a corpus is counted. From this joint probability of triples, one can condition upon the first two words, to get a conditional probability of the third. This makes the Markov assumption that the next state depends only on the current state, and that that state can be described by the previous two words. Under this assumption Eq. (3.2) becomes:

$$P(w^i \mid w^{i-1}, \ldots, w^1) = P(w^i \mid w^{i-1}, w^{i-2}) \tag{3.3}$$

More generally, one can use an n-gram language model where for any value of n, this is simply a matter of defining the Markov state to contain different numbers of words.

This Markov assumption is, of-course, an approximation. In the previous example, a trigram language model finds $P(w^i \mid \text{hot, a})$. It can be seen that the approximation has lost key information. Based only on the previous 2 words the next word w^i could now reasonably be day, but the sentence: For lunch I will have a hot day makes no sense. However, the Markov assumption in using n-grams is required in order to make the problem tractable – otherwise an unbounded amount of information would need to be stored.

A key issue with n-gram language models is that there exists a data-sparsity problem which causes issues in training them. Particularly for larger values of n. Most combinations of words occur very rarely (Ha et al. 2009). It is thus hard to estimate their occurrence probability. Combinations of words that do not occur in the corpus are naturally given a probability of zero. This is unlikely to be true though – it is simply a matter of rare phrases never occurring in a finite corpus. Several approaches have been taken to handle this. The simplest is add-one smoothing which adds an extra "fake" observation to every combination of terms. In common use are various back-off methods (Katz 1987; Kneser and Hermann 1995) which use the bigram probabilities to estimate the probabilities of unseen trigrams (and so forth for other n-grams.). However, these methods are merely clever statistical tricks – ways to reassign probability mass to leave some left-over for unseen cases. Back-off is smarter than add-one smoothing, as it portions the probability fairly based on the $(n-1)$-gram probability. Better still would be a method which can learn to see the common-role of words (Brown et al. 1992). By looking at the fragment: For lunch I want a hot, any reader knows that the next word is most likely going to be a food. We know this for the same reason we know the next word in For elevenses I had a cold . . . is also going to be a food. Even though elevenses is a vary rare word, we know from the context that it is a meal (more on this later), and we know it shares other traits with meals, and similarly have / had, and hot / cold. These traits influence the words that can occur after them. Hard-clustering words into groups is nontrivial, particularly given words having multiple meanings, and subtle differences in use. Thus the motivation is for a language modeling method which makes use of these shared properties of the words, but considers them in a flexible soft way. This motivates the need for representations which hold such linguistic information. Such representations must be discoverable from the corpus, as it is beyond reasonable to effectively hard-code suitable feature extractors. This is exactly the kind of task which a neural network achieves implicitly in its internal representations.

An extended look at classical techniques in statistical language modelling can be found in Goodman (2001).

3.1.1 The Neural Probabilistic Language Model

Bengio et al. (2003) present a method that uses a neural network to create a language model. In doing so it implicitly learns the crucial traits of words, during training. The core mechanism that allowed this was using an embedding or loop-up layer for the input.

Lookup word embeddings: Hashmap or Array?
The question is purely one of implementation. For purposes of the theory, it does not matter if the implementation is using a String to Vector dictionary (e.g. a hashmap), or a 2D array from which a column is indexed-out (sliced-from) via an integer index representing the word. In the tokenization of the source text, it is common to transform all the words into integers, so as to save memory, especially if string interning is not in use. At that point it makes sense to work with an array. For our notational purposes in this book, we will treat the word w^i as if it were an integer index, though thinking of it as a string index into a hashmap changes little in the logic.

3.1.1.1 Simplified Model Considered with Input Embeddings
To understand the neural probabilistic language model, let's first consider a simplified neural trigram language model. This model is a simplification of the model introduced by Bengio et al. (2003). It follows the same principles, and highlights the most important idea in neural language representations. This is that of training a vector representation of a word using a lookup table to map a discrete scalar word to a continuous-space vector which becomes the first layer of the network.

The neural trigram probabilistic network is defined by:

$$P(w^i \mid w^{i-1}, w^{i-2}) = \text{smax}\left(V\varphi\left(U\left[C_{:,w^{i-1}}; C_{:,w^{i-2}}\right] + \tilde{b}\right) + \tilde{k}\right) \qquad (3.4)$$

where U, V, $\tilde{b}$, $\tilde{k}$ are the weight matrices and biases of the network. The matrix C defines the embedding table, from which the word embeddings, $C_{:,w^{i-1}}$ and $C_{:,w^{i-2}}$, representing the previous two words (w^{i-1} and w^{i-2}) are retrieved. The network is shown in Fig. 3.2.

$C_{:,w^i}$ not $C_{:,i}$
Note that here we use the word w^i as the index to lookup the word embeddings. i is the index of the word index in the corpus. That is to say that if the ith word, and the jth word are the same: i.e $w^i = w^j$, then they will index out the same vector from C. $w^i = w^j \implies C_{:,w^i} = C_{:,w^j}$.

Fig. 3.2 The Neural Trigram
Language Model

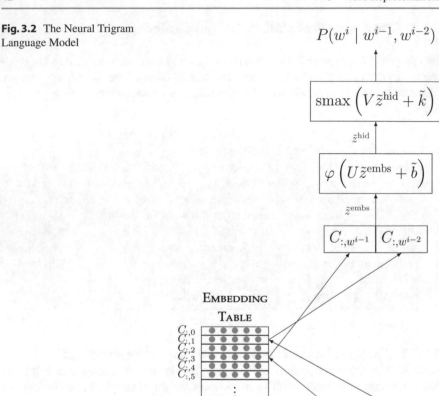

In the neural trigram language model, each of the previous two words is used to look-up a vector from the embedding matrix. These are then concatenated to give a dense, continuous-space input to the above hidden layer. The output layer is a softmax layer, it gives the probabilities for each word in the vocabulary, such that $\hat{y}_{w^i} = P(w^i \mid w^{i-1}, w^{i-2})$. Thus producing a useful language model.

The word embeddings are trained, just like any other parameter of the network (i.e. the other weights and biases) via gradient descent. An effect of this is that the embeddings of words which predict the same future word will be adjusted to be nearer to each other in the vector space. The hidden layer learns to associate information with regions of the embedding space, as the whole network (and every layer) is a continuous function. This effectively allows for information sharing between words. If two word's vectors are close together because they mostly predict the same future words, then that area of the embedding space is associated with predicting those words. If words a and b often occur as the word prior to some similar set of words $(w, x, y, \ldots)$ in the training set and word b also often occurs in the training set before word z, but (by chance) a never does, then this neural language model will predict that z is likely to occur after a. Where-as an n-gram language model would not. This is because a and b have similar embeddings, due to predicting a similar set of words. The model has learnt common features about these words implicitly from how they are used, and can use those to make better predictions. These features are stored in the embeddings which are looked up during the input.

3.1.1.2 Simplified Model Considered with Input and Output Embeddings

We can actually reinterpret the softmax output layer as also having embeddings. An alternative but equivalent diagram is shown in Fig. 3.3.

> **Consider, that the matrix product of a row vector with a column vector is the dot product**
> $V_{w_i,:}\tilde{z}^{\text{hid}}$ can be seen as computing the dot product between the output embedding for w_i and the hidden layer representation of the prior words/context (w^{i-1} and w^{i-2} in this case) in the form of $\tilde{z}^{\text{hid}}$. This leads to an alternate interpretation of the whole process as minimising the dot-product distance between the output embedding and the context representation This is particularly relevant for the skip-gram model discussed in Sect. 3.2.2 (with just one input word and no hidden layer).

The final layer of the neural trigram language model can be rewritten per each index corresponding to a possible next word (w^i):

$$\text{smax}(V\tilde{z}^{\text{hid}} + \tilde{k})_{w^i} = \frac{\exp\left(V_{w^i,:}\tilde{z}^{\text{hid}} + \tilde{k}_{w^i}\right)}{\sum_{\forall j} \exp\left(V_{j,:}\tilde{z}^{\text{hid}} + \tilde{k}_j\right)} \tag{3.5}$$

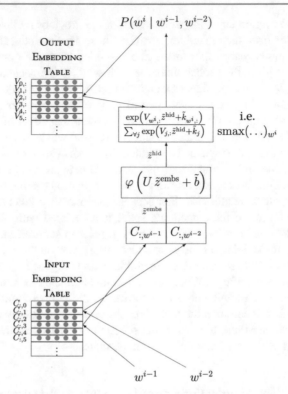

Fig. 3.3 Neural Trigram Language Model as considered with output embeddings. This is mathematically identical to Fig. 3.2

In this formulation, we have $V_{w_i,:}$ as the output embedding for w^i. As we considered $C_{:,w_i}$ as its input embedding.

3.1.1.3 Bayes-Like Reformulation
When we consider the model with output embeddings, it is natural to also consider it under the light of the Bayes-like reformulation from Sect. 1.5.1.1:

$$P(Y=i \mid Z=\tilde{z}) = \frac{R(Z=\tilde{z} \mid Y=i)\,R(Y=i)}{\sum_{\forall j} R(Z=\tilde{z} \mid Y=j)\,R(Y=j)} \quad (3.6)$$

which in this case is:

$$P(w^i \mid w^{i-1}, w^{i-2}) =$$
$$\frac{R(Z=\tilde{z}^{\text{hid}} \mid W^i=w^i)\,R(W^i=w^i)}{\sum_{\forall v \in \mathbb{V}} R(Z = \tilde{z}^{\text{hid}} \mid W^i=v)\,R(W^i=v)} \quad (3.7)$$

where $\sum_{\forall v \in \mathbb{V}}$ is summing over every possible word v from the vocabulary $\mathbb{V}$, which does include the case $v = w^i$.

Notice the term:

$$\frac{R(W^i = w^i)}{\sum_{\forall v \in \mathbb{V}} R(W^i = v)} = \frac{\exp\left(\tilde{k}_{w^i}\right)}{\sum_{\forall v \in \mathbb{V}} \exp\left(\tilde{k}_v\right)} \tag{3.8}$$

$$= \frac{1}{\sum_{\forall v \in \mathbb{V}} \exp\left(\tilde{k}_v - \tilde{k}_{w^i}\right)} \tag{3.9}$$

The term $R(W^i = w^i) = \exp(\tilde{k}_{w^i})$ is linked to the unigram word probabilities: $P(Y = y)$. If $\mathbb{E}(R(Z \mid W_i) = 1$ then the optimal value for $\tilde{k}$ would be given by the log unigram probabilities: $k_{w^i} = \log P(W^i = w^i)$. This condition is equivalent to if $\mathbb{E}(V \tilde{z}^{\text{hid}}) = 0$. Given that V is normally[1] initialized as a zero mean Gaussian, this condition is at least initially true. This suggests, interestingly, that we can pre-determine good initial values for the output bias $\tilde{k}$ using the log of the unigram probabilities. In practice this is not required, as it is learnt rapidly by the network during training.

3.1.1.4   The Neural Probabilistic Language Model

Bengio et al. (2003) derived a more advanced version of the neural language model discussed above. Rather than being a trigram language model, it is an n-gram language model, where n is a hyper-parameter of the model. The knowledge sharing allows the data-sparsity issues to be ameliorated, thus allowing for a larger n than in traditional n-gram language models. Bengio et al. (2003) investigated values for 2, 4 and 5 prior words (i.e. a trigram, 5-gram and 6-gram model). The network used in their work was marginally more complex than the trigram neural language model. As shown in Fig. 3.4, it features a layer-bypass connection. For n prior words, the model is described by:

$$\begin{aligned} P(w^i \mid w^{i-1}, \dots, w^{i-n}) = \text{smax}(&+ V \, \varphi\left(U^{\text{hid}}\left[C_{:,w^{i-1}}; \dots; C_{:,w^{i-n}}\right] + \tilde{b}\right) \\ &+ U^{\text{bypass}}\left[C_{:,w^{i-1}}; \dots; C_{:,w^{i-n}}\right] \\ &+ \tilde{k})_{w^i} \end{aligned} \tag{3.10}$$

The layer-bypass is a connivance to aid in the learning. It allows the input to directly affect the output without being mediated by the shared hidden layer. This layer-bypass is an unusual feature, not present in future works deriving from this, such as Schwenk (2004). Though in general it is not an unheard of technique in neural network machine learning.

---

[1] No pun intended.

Fig. 3.4 Neural Probabilistic
Language Model

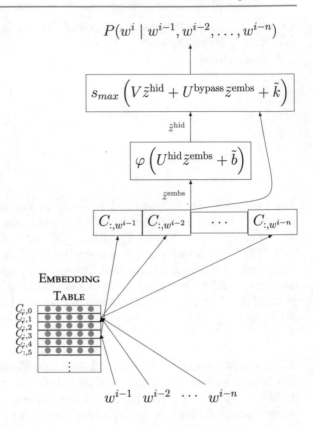

$$P(w^i \mid w^{i-1}, w^{i-2}, \ldots, w^{i-n})$$

$$s_{max}\left(V\tilde{z}^{\text{hid}} + U^{\text{bypass}}\tilde{z}^{\text{embs}} + \tilde{k}\right)$$

$\tilde{z}^{\text{hid}}$

$$\varphi\left(U^{\text{hid}}\tilde{z}^{\text{embs}} + \tilde{b}\right)$$

$\tilde{z}^{\text{embs}}$

$C_{:,w^{i-1}}$ $C_{:,w^{i-2}}$ $\cdots$ $C_{:,w^{i-n}}$

EMBEDDING

TABLE

$C_{:,0}$
$C_{:,1}$
$C_{:,2}$
$C_{:,3}$
$C_{:,4}$
$C_{:,5}$

w^{i-1} w^{i-2} $\cdots$ w^{i-n}

Input vocabulary and output vocabulary do not have to be the same
Schwenk (2004) suggests using only a subset of the vocabulary as options
for the output, while allowing the full vocabulary in the input space – with a
fall-back to classical language models for the missed words. This decreases
the size of the softmax output layer, which substantially decreases the time
taken to train or evaluate the network. As a speed-up technique this is now
eclipsed by hierarchical softmax and negative sampling discussed in Sect. 3.4.
The notion of a different input and output vocabulary though remains important
for word-sense embeddings as will be discussed in this chapter.

This is the network which begins the notions of using neural networks with vector
representations of words. Bengio et al. focused on the use of the of sliding window
of previous words – much like the traditional n-grams. At each time-step the window
is advanced forward and the next is word predicted based on the shifted context of
prior words. This is of-course exactly identical to extracting all n-grams from the
corpus and using those as the training data. They very briefly mention that an RNN
could be used in its place.

3.1.2 RNN Language Models

In Mikolov et al. (2010) an RNN is used for language modelling, as shown in Fig. 3.5. Using the terminology of Chap. 2, this is an encoder RNN, made using Basic Recurrent Units. Using an RNN eliminates the Markov assumption of a finite window of prior words forming the state. Instead, the state is learned, and stored in the state component of the RUs.

No Bias?

It should be noticed that Eqs. (3.11) and (3.12) are missing the bias terms. This is not commented on in Mikolov et al. (2010). But in the corresponding chapter of Mikolov's thesis (Tomas 2012), it is explicitly noted that biases were not used in the network as it was not found that they gave a significant improvement to the result. This is perhaps surprising, particularly in the output softmax layer given the very unbalanced class (unigram) frequencies.

In the papers for several of Mikolov's other works, including those for skip-gram and CBOW discussed in Sect. 3.2, the bias terms are also excluded. We have matched those equations here. We do note though, that it is likely that many publicly available implementations of these algorithms would include the bias term: due either to a less close reading of the papers, or to the assumption that the equations are given in *design matrix* form: where the bias is not treated as a separate term to the weights, and the input is padded with an extra 1. We do not think this is at all problematic.

We discuss this further for the case of hierarchical softmax in Sect. 3.4.1, where the level is a proxy for the unigram frequency – and thus for the bias.

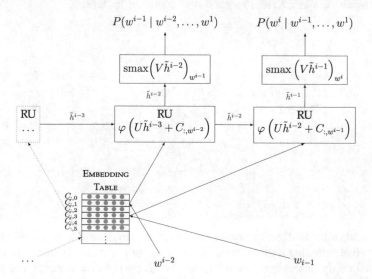

Fig. 3.5 RNN Language Model. The RU equation shown is the basic RU used in Mikolov et al. (2010). It can be substituted for a LSTM RU or an GRU as was done in Sundermeyer et al. (2012); Jozefowicz et al. (2015), with appropriate changes

This state $\tilde{h}_i$ being the hidden state (and output as this is a basic RU) from the i time-step. The ith time-step takes as its input the ith word. As usual this hidden layer was an input to the hidden-layer at the next time-step, as well as to the output softmax.

$$\tilde{h}^i = \varphi \left(U\tilde{h}^{i-1} + C_{:,w_{i-1}} \right) \tag{3.11}$$

$$P(w^i \mid w^{i-1}, \ldots, w^1) = \text{smax} \left(V\tilde{h}^{i-1} \right)_{w^i} \tag{3.12}$$

Rather than using a basic RU, a more advanced RNN such as a LSTM or GRU-based network can be used. This was done by Sundermeyer et al. (2012) and Jozefowicz et al. (2015), both of whom found that the more advanced networks gave significantly better results.

3.2 Acausal Language Modeling

The step beyond a normal language model, which uses the prior words to predict the next word, is what we will term acausal language modelling. Here we use the word acausal in the signal processing sense. It is also sometimes called contextual language modelling, as the whole context is used, not just the prior context. The task here is to predict a missing word, using the words that precede it, as well as the words that come after it.

As it is acausal it cannot be implemented in a real-time system, and for many tasks this renders it less, directly, useful than a normal language model. However, it is very useful as a task to learn a good representation for words.

Are CBOW & Skip-Gram Neural Networks?
It is sometimes asserted that these models are not in-fact neural networks at all. This assertion is often based on their lack of a traditional hidden-layer, and similarities in form to several other mathematical models (discussed in Sect. 3.3). This distinction is purely academic though. Any toolkit that can handle the prior discussed neural network models can be used to implement CBOW and Skip-Gram, more simply than using a non-neural network focused optimiser.

It also should be noted that embedding lookup is functionally an unusual hidden layer – this becomes obvious when considering the lookup as an one-hot product. Though it does lack a non-linear activation function.

The several of the works discussed in this section also feature hierarchical softmax and negative sampling methods as alternative output methods. As these are complicated and easily misunderstood topics they are discussed in a more tutorial fashion in Sect. 3.4. This section will focus just on the language model logic; and assume the output is a normal softmax layer.

3.2.1 Continuous Bag of Words

The continuous bag of words (CBOW) method was introduced by Mikolov et al. (2013a). In truth, this is not particularly similar to bag of words at all. No more so than any other word representation that does not have regard for order of the context words (e.g. skip-gram, and GloVe).

The CBOW model takes as its input a context window surrounding a central skipped word, and tries to predict the word that it skipped over. It is very similar to earlier discussed neural language models, except that the window is on both sides. It also does not have any non-linearities; and the only hidden layer is the embedding layer.

For a context window of width n words – i.e. $\frac{n}{2}$ words to either side, of the target word w^i, the CBOW model is defined by:

$$P(w^i \mid w^{i-\frac{n}{2}}, \ldots, w^{i-1}, w^{i+1}, \ldots, w^{i+\frac{n}{2}})$$

$$= \mathrm{smax} \left(V \sum_{\substack{j=i+1 \\ }}^{j=\frac{n}{2}} \left(C_{:,w^{i-j}} + C_{:,w^{i+j}} \right) \right)_{w^i} \tag{3.13}$$

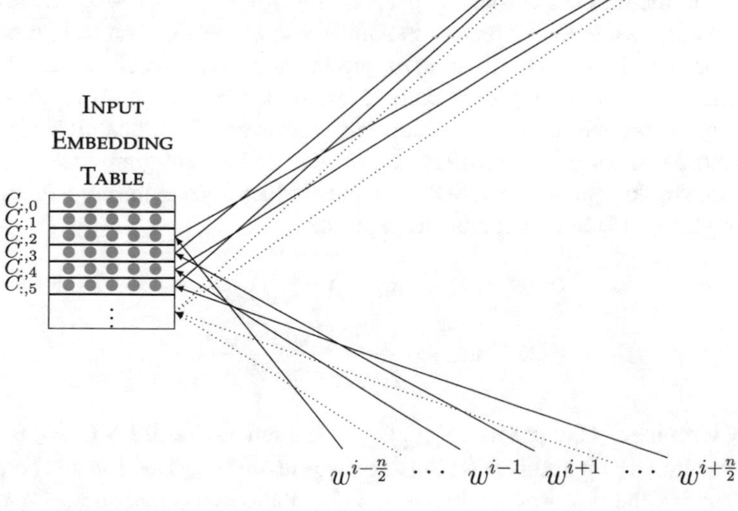

Fig. 3.6 CBOW Language Model

This is shown in diagrammatic form in Fig. 3.6. By optimising across a training dataset, useful word embeddings are found, just like in the normal language model approaches.

3.2.2   Skip-Gram

Skip-gram naming
In different publications this model may be called skipgram, skip-gram, skip-ngram, skip gram etc. Further, it may be called word2vec after the publicly released implementation of the algorithm. Though the word2vec software can also be used for CBOW, so sometimes it can refer to CBOW.

The converse of CBOW is the skip-grams model (Mikolov et al. 2013a). In this model, the central word is used to predict the words in the context.

The model itself is single word input, and its output is a softmax for the probability of each word in the vocabulary occurring in the context of the input word. This can be indexed to get the individual probability of a given word occurring as usual for a language model. So for input word w^i the probability of w^j occurring in its context is given by:

$$P(w^j \mid w^i) = \text{smax}\left(V\, C_{:,w^i}\right)_{w^j} \tag{3.14}$$

This is shown in Fig. 3.7.

The goal, is to maximise the probabilities of all the observed outputs that actually *do* occur in its context. This is done, as in CBOW by defining a window for the context of a word in the training corpus, $(i - \frac{n}{2}, \ldots, i - 1, i + i, \ldots, i + \frac{n}{2})$. It should be understood that while this is presented similarly to a classification task, there is no expectation that the model will actually predict the correct result, given that even during training there are multiple correct results. It is a regression to an accurate estimate of the probabilities of co-occurrence (this is true for probabilistic language models more generally, but is particularly obvious in the skip-gram case).

Note that in skip-gram, like CBOW, the only hidden layer is the embedding layer. Rewriting Eq. (3.14) in output embedding form:

$$P(w^j \mid w^i) = \text{smax}\left(V\, C_{:,w^i}\right)_{w^j} \tag{3.15}$$

$$P(w^j \mid w^i) = \frac{\exp\left(V_{w^j,:} C_{:,w^i}\right)}{\sum_{\forall v \in \mathbb{V}} \exp\left(V_{v,:} C_{:,v}\right)} \tag{3.16}$$

The key term here is the product $V_{w^j,:}\, C_{:,w^i}$. The remainder of Eq. (3.16) is to normalise this into a probability. Maximising the probability $P(w^j \mid w^i)$ is equivalent to maximising the dot produce between $V_{w^j,:}$, the output embedding for w^j and $C_{:,w^i}$ the input embedding for w^i. This is to say that the skip-gram probability is

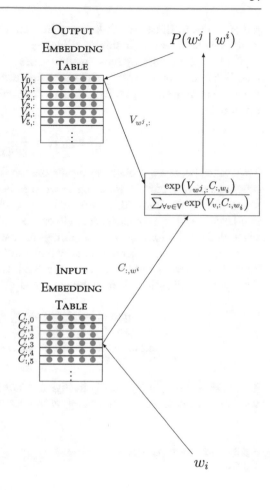

Fig. 3.7 Skip-gram language Language Model. Note that the probability $P(w^j \mid w^i)$ is optimised during training for every w^j in a window around the central word w^i. Note that the final layer in this diagram is just a softmax layer, written in in output embedding form

maximised when the angular difference between the input embedding for a word, and the output embeddings for its co-occurring words is minimised. The dot-product is a measure of vector similarity – closely related to the cosine similarity.

Skip-gram is much more commonly used than CBOW.

3.2.3 Analogy Tasks

One of the most notable features of word embeddings is their ability to be used to express analogies using linear algebra. These tasks are keyed around answering the question: b is to a, as what is to c? For example, a semantic analogy would be answering that Aunt is to Uncle as King is to Queen. A syntactic analogy would be answering that King is to Kings as Queen is to Queens. The latest and largest analogy test set is presented by Gladkova et al. (2016), which evaluates embeddings on 40 subcategories of knowledge. Analogy completion is not a practical task, but rather serves to illustrate the kinds of information being captured, and the way in which it is represented (in this case linearly).

The analogies work by relating similarities of differences between the word vectors. When evaluating word similarity using using word embeddings a number of measures can be employed. By far the cosine similarity is the most common. This is given by

$$\text{sim}(\tilde{u}, \tilde{v}) = \frac{\tilde{u} \cdot \tilde{v}}{\|\tilde{u}\| \, \|\tilde{v}\|} \tag{3.17}$$

This value becomes higher the closer the word embedding $\tilde{u}$ and $\tilde{v}$ are to each other, ignoring vector magnitude. For word embeddings that are working well, then words with closer embeddings should have correspondingly greater similarity. This similarity could be syntactic, semantic or other. The analogy tasks can help identify what kinds of similarities the embeddings are capturing.

Using the similarity scores, a ranking of words to complete the analogy is found. To find the correct word for d in: d is to c as b is to a the following is computed using the table of embeddings C over the vocabulary $\mathbb{V}$:

$$\operatorname*{argmax}_{\forall d \in \mathbb{V}} \; \text{sim}(C_{:}, d - C_{:}, c, C_{:}, a - C_{:}, b) \tag{3.18}$$

$$\text{i.e} \operatorname*{argmax}_{\forall d \in \mathbb{V}} \; \text{sim}(C_{:}, d, \; C_{:}, a - C_{:}, b + C_{:}, c) \tag{3.19}$$

This is shown diagrammatically in Figs. 3.8 and 3.9. Sets of embeddings where the vector displacement between analogy terms are more consistent score better.

Fig. 3.8 Example of analogy algebra

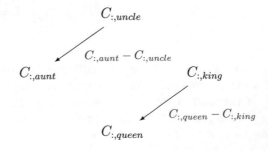

Fig. 3.9 Vectors involved in analogy ranking tasks, this may help to understand the math in Eq. (3.19)

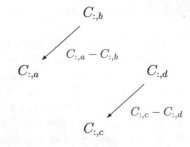

Initial results in Mikolov et al. (2013c) were relatively poor, but the surprising finding was that this worked at all. Mikolov et al. (2013a) found that CBOW performed poorly for semantic tasks, but comparatively well for syntactic tasks; skip-gram performed comparatively well for both, though not quite as good in the syntactic tasks as CBOW. Subsequent results found in Pennington et al. (2014) were significantly better again for both.

Analogy Tasks uncover prejudice in corpora
Bolukbasi et al. (2016) and Caliskan et al. (2017) use analogy tasks, and related variant formulations to find troubling associations between words, such as Bolukbasi et. al's titular `Man` is to `Computer Programmer`, as `Woman` is to `Homemaker`. Finding these relationships in the embedding space, indicated that they are present in the training corpus, which in turn shows their prevalence in society at large. It has been observed that machine learning can be a very good mirror upon society.

3.3   Co-location Factorisation

3.3.1   GloVe

Implementing GloVe
To implement GloVe in any technical programming language with good support for optimisation is quiet easy, as it is formed into a pure optimization problem. It is also easy to do in a neural network framework, as these always include an optimiser. Though unlike in normal neural network training there are no discrete training cases, just the global co-occurrence statistics.

Skip-gram, like all probabilistic language models, is a intrinsically prediction-based method. It is effectively optimising a neutral network to predict which words will co-occur in the with in the range of given by the context window width. That optimisation is carried out per-context window, that is to say the network is updated based on the local co-occurrences. In Pennington et al. (2014) the authors show that if one were to change that optimisation to be global over all co-occurrences, then the optimisation criteria becomes minimising the cross-entropy between the true co-occurrence probabilities, and the value of the embedding product, with the cross entropy measure being weighted by the frequency of the occurrence of the word. That is to say if skip-gram were optimised globally it would be equivalent to minimising:

$$Loss = - \sum_{\forall w^i \in \mathbb{V}} \sum_{\forall w^j \in \mathbb{V}} X_{w^i, w^j} P(w^j \mid w^i) \log(V_{w^j,:} C_{:,w^i}) \qquad (3.20)$$

for $\mathbb{V}$ being the vocabulary and for X being the a matrix of the true co-occurrence counts, (such that X_{w^i,w^j} is the number of times words w^i and w^j co-occur), and for P being the predicted probability output by the skip-gram.

Minimising this cross-entropy efficiently means factorising the true co-occurrence matrix X, into the input and output embedding matrices C and V, under a particular set of weightings given by the cross entropy measure.

Pennington et al. (2014) propose an approach based on this idea. For each word co-occurrence of w^i and w^j in the vocabulary: they attempt to find optimal values for the embedding tables C, V and the per word biases $\tilde{b}$, $\tilde{k}$ such that the function $s(w^i, w^j)$ (below) expresses an approximate log-likelihood of w^i and w^j.

$$\text{optimise} \quad s(w^i, w^j) \quad = V_{w^j,:} \cdot C_{:,w^i} + \tilde{b}_{w^i} + \tilde{k}_{w^j} \tag{3.21}$$

$$\text{such that} \quad s(w^i, w^j) \quad \approx \log(X_{w^i,w^j}) \tag{3.22}$$

This is done via the minimisation of

$$Loss = -\sum_{\forall w^i}\sum_{\forall w^j} f(X_{w^i,w^j}) \left(s(w^i, w^j) - \log(X_{w^i,w^j}) \right) \tag{3.23}$$

Where $f(x)$ is a weighing between 0 and 1 given by:

$$f(x) = \begin{cases} \left(\frac{x}{100}\right)^{0.75} & x < 100 \\ 1 & \text{otherwise} \end{cases} \tag{3.24}$$

This can be considered as a saturating variant of the effective weighing of skip-gram being X_{w^i,w^j}.

While GloVe out-performed skip-gram in initial tests subsequent more extensive testing in Levy et al. (2015) with more tuned parameters, found that skip-gram marginally out-performed GloVe on all tasks.

> **Key Factors Mentioned as Asides**
> There is an interesting pattern of factors being considered as not part of the core algorithm. We have continued this in the side-notes of this section; with the preceding notes on Distance weighting and subsampling. While the original papers consider these as unimportant to the main thrust of the algorithms (Levy et al. 2015) found them to be crucial hyper-parameters.

Distance weighted co-occurrence and dynamic window sizing
When training skip-gram and CBOW, Mikolov et al. used dynamic window sizing. This meant that if the specified window size was n, in any given training case being considered the actual window size was determined as a random number between 0 and n.

Pennington et al. achieve a similar effect by weighting co-occurrences within a window with inverse proportion to the distance between the word. That is to say if w^i and w^j occur in the same window (i.e. $|i - j| < n$), then rather than contributing 1 to the entry in the co-occurrence count X_{w^i, w^j}, they contribute $\dfrac{1}{|i - j|}$.

Subsampling, and weight saturation
Skip-gram and CBOW models use a method called subsampling to decrease the effect of common words. The subsampling method is to randomly discard words from training windows based on their unigram frequency. This is closely related to the saturation of the co-occurrence weights as calculated by $f(X)$ used by GloVe. Averaged over all training cases the effect is nearly the same.

3.3.2 Further Equivalence of Co-location Prediction to Factorisation

GloVe highlights the relationship between the co-located word prediction neural network models, and the more traditional non-negative matrix factorization of co-location counts used in topic modeling. Very similar properties were also explored for skip-grams with negative sampling in Levy and Yoav (2014) and in Li et al. (2015) with more direct mathematical equivalence to weighed co-occurrence matrix factorisation; Later, Cotterell et al. (2017) showed the equivalence to exponential principal component analysis (PCA). Landgraf and Jeremy (2017) goes on to extend this to show that it is a weighted logistic PCA, which is a special case of the exponential PCA. Many works exist in this area now.

3.3.3 Conclusion

We have now concluded that neural predictive co-location models are functionally very similar to matrix factorisation of co-location counts with suitable weightings, and suitable similarity metrics. One might now suggest a variety of word embeddings to be created from a variety of different matrix factorisations with different weightings and constraints. Traditionally large matrix factorisations have significant problems in terms of computational time and memory usage. A common solution to this, in applied mathematics, is to handle the factorisation using an iterative optimisation procedure. Training a neural network, such as skip-gram, is indeed just such an iterative optimisation procedure.

3.4   Hierarchical Softmax and Negative Sampling

Hierarchical softmax, and negative sampling are effectively alternative output layers which are computationally cheaper to evaluate than regular softmax. They are powerful methods which pragmatically allow for large speed-up in any task which involves outputting very large classification probabilities – such as language modelling.

3.4.1   Hierarchical Softmax

Hierarchical softmax was first presented in Morin and Bengio (2005). Its recent use was popularised by Mikolov et al. (2013a), where words are placed as leaves in a Huffman tree, with their depth determined by their frequency.

> **SemHuff**
> It can be noted that the Huffman encoding scheme specifies only the depth of a given word in the tree. It does not specify the order. Yang et al. (2016) make use of the BlossomV algorithm (Kolmogorov 2009) to pair the nodes on each layer according to their similarity. They found that on the language modelling task this improved performance, in the way one would expect. They used a lexical resource to determine similarity, however noted that a prior trained word-embedding model could be used to define similarity instead – the new encoding can then be used to define a new model which will find new (hopefully better) embeddings. This is similar to the original method used by (Morin and Bengio 2005), but only using the similarity measure for reordering nodes at the same depth, after the depth is decided by Huffman encoding. In our own experimentation, when applying it to other tasks, we did not see large improvements. It is nevertheless a very interesting idea, and quite fun to implement and observe the results.

One of the most expensive parts of training and using a neural language model is to calculate the final softmax layer output. This is because the softmax denominator includes terms for each word in the vocabulary. Even if only one word's probability is to be calculated, one denominator term per word in the vocabulary must be evaluated.

In hierarchical softmax, each word (output choice), is considered as a leaf on a binary tree. Each level of the tree roughly halves the space of the output words to be considered. The final level to be evaluated for a given word contains the word's leaf-node and another branch, which may be a leaf-node for another word, or a deeper sub-tree

The tree is normally a Huffman tree (Huffman 1952), as was found to be effective by Mikolov et al. (2013a). This means that for each word w^i, the word's depth (i.e its code's length) $l(w^i)$ is such that over all words: $\sum_{\forall w^j \in \mathbb{V}} P(w^j) \times l(w^j)$ is minimised. Where $P(w^i)$ is word w^i's unigram probability, and $\mathbb{V}$ is the vocabulary. The approximate solution to this is that $l(w^i) \approx -\log_2(P(w^i))$. From the tree, each word can be assign a code in the usual way, with 0 for example representing taking one branch, and 1 representing the other. Each point in the code corresponds to a node in the binary tree, which has decision tied to it. This code is used to transform the large multinomial softmax classification into a series of binary logistic classifications. It is important to understand that the layers in the tree are not layers of the neural network in the normal sense – the layers of the tree do not have an output that is used as the input to another. The layers of the tree are rather subsets of the neurons on the output layer, with a relationship imparted on them.

It was noted by Mikolov et al. (2013a), that for vocabulary $\mathbb{V}$:

- Using normal softmax would require each evaluation to perform $|\mathbb{V}|$ operations.
- Using hierarchical softmax with a balanced tree, would mean the expected number of operations across all words would be $\log_2(|\mathbb{V}|)$.
- Using a Huffman tree gives the expected number of operations $\sum_{\forall w^j \in \mathbb{V}} -P(w^j)$ $\log_2(P(w^i)) = H(\mathbb{V})$, where $H(\mathbb{V})$ is the unigram entropy of words in the training corpus.

The worse case value for the entropy is $\log_2(|\mathbb{V}|)$. In-fact Huffman encoding is provably optimal in this way. As such this is the minimal number of operations required in the average case.

3.4.1.1   An Incredibly Gentle Introduction to Hierarchical Softmax

In this section, for brevity, we will ignore the bias component of each decision at each node. It can either be handled nearly identically to the weight; or the matrix can be written in *design matrix form* with an implicitly appended column of ones; or it can even be ignored in the implementation (as was done in Mikolov et al. 2013a). The reasoning for being able to ignore it is that the bias in normal softmax encodes unigram probability information; in hierarchical softmax, when used with the common Huffman encoding, its the tree's depth in tree encodes its unigram probability. In this case, not using a bias would at most cause an error proportionate to 2^{-k}, where k is the smallest integer such that $2^{-k} > P(w^i)$.

First Consider a Binary Tree with just 1 Layer and 2 Leaves
The leaves are n^{00} and n^{01}, each of these leaf nodes corresponds to a word from the vocabulary, which has size two, for this toy example. This is shown in Fig. 3.10.

From the initial root which we call n^0, we can go to either node n^{00} or node n^{01}, based on the input from the layer below which we will call $\tilde{z}$.

Fig. 3.10 Tree for 2 words

Here we write n^{01} to represent the event of the first non-root node being the branch given by following left branch, while n^{01} being to follow the right branch. (The order within the same level is arbitrary in any-case, but for our visualisation purposes we'll used this convention.)

We are naming the root node as a notation convenience so we can talk about the decision made at n^0. Note that $P(n^0) = 1$, as all words include the root-node on their path.

We wish to know the probability of the next node being the left node (i.e. $P(n^{00} \mid \tilde{z})$) or the right-node (i.e. $P(n^{01} \mid \tilde{z})$). As these are leaf nodes, the prediction either equivalent to the prediction of one or the other of the two words in our vocabulary.

We could represent the decision with a softmax with two outputs. However, since it is a binary decision, we do not need a softmax, we can just use a sigmoid.

$$P(n^{01} \mid \tilde{z}) = 1 - P(n^{00} \mid \tilde{z}) \tag{3.25}$$

The weight matrix for a sigmoid layer has a number of columns governed by the number of outputs. As there is only one output, it is just a row vector. We are going to index it out of a matrix V. For the notation, we will use index 0 as it is associated with the decision at node n^0. Thus we call it $V_{0,:}$.

$$P(n^{00} \mid \tilde{z}) = \sigma(V_{0,:}\tilde{z}) \tag{3.26}$$

$$P(n^{01} \mid \tilde{z}) = 1 - \sigma(V_{0,:}\tilde{z}) \tag{3.27}$$

Note that for the sigmoid function: $1 - \sigma(x) = \sigma(-x)$. Allowing the formulation to be written:

$$P(n^{01} \mid \tilde{z}) = \sigma(-V_{0,:}\tilde{z}) \tag{3.28}$$

thus

$$P(n^{0i} \mid \tilde{z}) = \sigma((-1)^i V_{0,:}\tilde{z}) \tag{3.29}$$

Noting that in Eq. (3.29), i is either 0 (with $-1^0 = 1$) or 1 (with $-1^1 = -1$)).

$V_{0,:}\tilde{z}$ is a dot product
We mentioned in the marginalia earlier, but just as an extra reminder: the matrix product of a row vector like $V_{0,:}$ with a (column) vector like $\tilde{z}$ is their vector dot product.

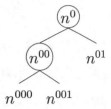

Fig. 3.11 Tree for 3 words

Now Consider 2 Layers with 3 Leaves

Consider a tree with nodes: n^0, n^{00}, n^{000}, n^{001}, n^{01}. The leaves are n^{000}, n^{001}, and n^{01}, each of which represents one of the 3 words from the vocabulary (Fig. 3.11).

From earlier we still have:

$$P(n^{00} \mid \tilde{z}) = \sigma(V_{0,:}\tilde{z}) \tag{3.30}$$

$$P(n^{01} \mid \tilde{z}) = \sigma(-V_{0,:}\tilde{z}) \tag{3.31}$$

We must now to calculate $P(n^{000} \mid \tilde{z})$. Another binary decision must be made at node n^{00}. The decision at n^{00} is to find out if the predicted next node is n^{000} or n^{001}. This decision is made, with the assumption that we have reached n^{00} already.

So the decision is defined by $P(n^{000} \mid z, n^{00})$ is given by:

$$P(n^{000} \mid \tilde{z}) = P(n^{000} \mid \tilde{z}, n^{00}) \, P(n^{00} \mid \tilde{z}) \tag{3.32}$$

$$P(n^{000} \mid \tilde{z}, n^{00}) = \sigma(V_{00,:}\tilde{z}) \tag{3.33}$$

$$P(n^{001} \mid \tilde{z}, n^{00}) = \sigma(-V_{00,:}\tilde{z}) \tag{3.34}$$

We can use the conditional probability chain rule to recombine to compute the three leaf nodes final probabilities.

$$P(n^{01} \mid \tilde{z}) = \sigma(-V_{0,:}\tilde{z}) \tag{3.35}$$

$$P(n^{000} \mid \tilde{z}) = \sigma(V_{00,:}\tilde{z})\sigma(V_{0,:}\tilde{z}) \tag{3.36}$$

$$P(n^{001} \mid \tilde{z}) = \sigma(-V_{00,:}\tilde{z})\sigma(V_{0,:}\tilde{z}) \tag{3.37}$$

Continuing this Logic

Using this system, we know that for a node encoded at position $[0t^1t^2t^3 \ldots t^L]$, e.g. $[010\ldots1]$, its probability can be found recursively as:

$$P(n^{0t^1\ldots t^L} \mid \tilde{z}) = P(n^{0t^1\ldots t^L} \mid \tilde{z}, n^{0t^1\ldots t^{L-1}}) \, P(n^{0t^1\ldots t^{L-1}} \mid \tilde{z}) \tag{3.38}$$

Thus:

$$P(n^{0t^1} \mid \tilde{z}) = \sigma \left((-1)^{t^1} V_{0,:}\tilde{z} \right) \tag{3.39}$$

$$P(n^{0t^1,t^2} \mid \tilde{z}, n^{0t^1}) = \sigma \left((-1)^{t^2} V_{0t^1,:}\tilde{z} \right) \tag{3.40}$$

$$P(n^{0t^1\dots t^i} \mid \tilde{z}, n^{0t^1\dots t^{i-1}}) = \sigma \left((-1)^{t^i} V_{0t^1\dots t^{i-1},:}\tilde{z} \right) \tag{3.41}$$

The conditional probability chain rule, is applied to get:

$$P(n^{0t^1\dots t^L} \mid \tilde{z}) = \prod_{i=1}^{i=L} \sigma \left((-1)^{t^i} V_{0t^1\dots t^{i-1},:}\tilde{z} \right) \tag{3.42}$$

3.4.1.2 Formulation

Combining multiplications
If one wants to find both $V_{00,:}\tilde{z}$ and $V_{0,:}\tilde{z}$, then this can be done using matrices simultaneously, thus potentially taking advantage of optimized matrix multiplication routines.

$$\begin{bmatrix} V_{0,:} \\ V_{00,:} \end{bmatrix} z = \begin{bmatrix} V_{0,:}\tilde{z} \\ V_{00,:}\tilde{z} \end{bmatrix}$$

Thus the whole product for all of the decisions can be written as $V\tilde{z}$. The problem then becomes indexing the relevant node rows.

However computing every single decision is beyond what is required for most uses: hierarchical softmax lets us only compute the decisions that are on the path to the word-leaf we which we wish to query. Computing all of them is beyond what is required.

Packing tree node elements into a matrix with fast indexing is a non-trivial problem. The details on optimising such multiplications and tree packing are beyond the scope of this book.

In general there may be very little scope here for optimisation, as on most hardware (and BLAS systems) matrix products with n columns, takes a similar amount of time to n vector dot products. As such storing the row vectors of V in a hashmap indexed by node-path, and looping over them as required may be more practical.

In languages/libraries with slow looping constructs (numpy, R, octave), where calling into suitable library routines is much faster, this may give some speed-up; but even here it is likely to be minor. The time may be better spent writing a C extension library to do this part of the program. Or learning to use a language with fast for loops (e.g. Julia 2014).

The formulation above is not the same as in other works. This subsection shows the final steps to reach the conventional form used in Mikolov et al. (2013b).

Here we have determined that the 0th/left branch represents the positive choice, and the other probability is defined in terms of this. It is equivalent to have the 1th/right branch representing the positive choice:

$$P(n^{0t^1...t^L} \mid \tilde{z}) = \prod_{i=1}^{i=L} \sigma \left((-1)^{t^i+1} V_{0t^1...t^{i-1},:} \tilde{z} \right) \tag{3.43}$$

or to allow it to vary per node: as in the formulation of Mikolov et al. (2013b). In that work they use $ch(n)$ to represent an arbitrary child node of the node n and use an indicator function $[\![a = b]\!] = \begin{cases} 1 & a = b \\ -1 & a \neq b \end{cases}$ such that they can write $[\![n^b = ch(n^a)]\!]$ which will be 1 if n^a is an arbitrary (but consistent) child of n^b, and 0 otherwise.

$$P(n^{0t^1...t^L} \mid \tilde{z}) =$$
$$\prod_{i=1}^{i=L} \sigma \left([\![n^{0t^1...t^i} = ch(n^{0t^1...t^{i-1}})]\!] V_{0t^1...t^{i-1},:} \tilde{z} \right) \tag{3.44}$$

There is no functional difference between the three formulations. Though the final one is perhaps a key reason for the difficulties in understanding the hierarchical softmax algorithm.

How does this relate to word vectors?
After the length of this section, one may have forgotten why we are doing this in the first place. Recall that CBOW, skip-gram and all other language modelling based word embedding methods are based around predicting $P(w^o \mid w^i, \ldots, w^j)$ for some words. For skip-gram that is just $P(w^o \mid w^i)$. The term $n^{0t^1...t^L}$ in $P(n^{0t^1...t^L} \mid z)$, just represents as a path through the tree to the leaf node which represents the word w^o. i.e $P(n^{0t^1...t^L} \mid z) = P(w^o \mid z)$. The output of the final hidden layer is z (i.e. the z is the input to the output layer) In normal language models z encodes all the information about what the model knows of predictions based on $w^i \ldots, w^j$. z is thus a proxy term in the conditional probability for those words. In skip-gram there is no hidden layer, and it is just $z = C_{:,w^i}$ proxying only for w_i, and the model defines its probability output by $P(w^o \mid w^i) = P(w^o \mid C_{:,w^i})$.

3.4.1.3   Loss Function

The gradient calculations
They are not fun. They never are for back-propagation. We recommend using a framework with automated differentiation, and/or performing gradient checks against a numerical differentiation tool (simple finite-differencing will do in a pinch).

Using normal softmax, during the training, the cross-entropy between the model's predictions and the ground truth as given in the training set is minimised. Cross entropy is given by

$$CE(P^\star, P) = \sum_{\forall w^i \in \mathbb{V}} \sum_{\forall z^j \in \mathbb{Z}} -P^\star(w^i \mid z^j) \log P(w^i \mid z^j) \qquad (3.45)$$

where $P^\star$ is the true distribution, and P is the approximate distribution given by our model (in other sections we have abused notation to use P for both). $\mathbb{Z}$ is the set of values that are input into the model, (or equivalently the values derived from them from lower layers) – Ithe context words in language modelling. $\mathbb{V}$ is the set of outputs, the vocabulary in language modeling. The training dataset $\mathcal{X}$ consists of pairs from $\mathbb{V} \times \mathbb{Z}$.

The true probabilities (from $P^\star$) are implicitly given by the frequency of the training pairs in the training dataset $\mathcal{X}$.

$$Loss = CE(P^\star, P) = \frac{1}{|\mathcal{X}|} \sum_{\forall (w^i, z^i) \in \mathcal{X}} -\log P(w^i \mid z^i) \qquad (3.46)$$

The intuitive understanding of this, is that we are maximising the probability estimate of all pairings which actually occur in the training set, proportionate to how often the occur. Note that the $\mathbb{Z}$ can be non-discrete values, as was the whole benefit of using embeddings, as discussed in Sect. 3.1.1.

This works identically for hierarchical softmax as for normal softmax. It is simply a matter of substituting in the (different) equations for P. Then applying back-propagation as usual.

3.4.2   Negative Sampling

Negative sampling was introduced in Mikolov et al. (2013b) as another method to speed up this problem. Much like hierarchical softmax in its purpose. However, negative sampling does not modify the network's output, but rather the loss function.

Negative Sampling is a simplification of Noise Contrast Estimation (Gutmann and Hyvärinen 2012). Unlike Noise Contrast Estimation (and unlike softmax), it does not in fact result in the model converging to the same output as if it were trained with softmax and cross-entropy loss. However the goal with these word embeddings is not to actually perform the language modelling task, but only to capture a high-quality vector representation of the words involved.

3.4.2.1 A Motivation of Negative Sampling

Recall from Sect. 3.2.2 that the (supposed) goal, is to estimate $P(w^j \mid w^i)$. In truth, the goal is just to get a good representation, but that is achieved via optimising the model to predict the words. In Sect. 3.2.2 we considered the representation of $P(w^j \mid w^i)$ as the w^jth element of the softmax output.

$$P(w^j \mid w^i) = \text{smax}(V\,C_{:,w^i})_{w^j} \tag{3.47}$$

$$P(w^j \mid w^i) = \frac{\exp\left(V_{w^j,:}C_{:,w^i}\right)}{\sum_{k=1}^{k=N} \exp\left(V_{k,:}C_{:,k}\right)} \tag{3.48}$$

> **Why is not using softmax wrong?**
> The notation abuse may be hiding just how bad it is to not use softmax.
> Recall that the true meaning of $P(w^j \mid w^i)$ is actually $P(W^j{=}w^j \mid W^i{=}w^i)$.
> By not using softmax, with its normalising denominator this means that:
> $\sum_{\forall w^j \in \mathbb{V}} P(w^j \mid w^i) \neq 1$ (except by coincidence).

This is not the only valid representation. One could use a sigmoid neuron for a direct answer to the co-location probability of w^j occurring near w^i. Though this would throw away the promise of the probability distribution to sum to one across all possible words that could be co-located with w^i. That promise could be enforced by other constraints during training, but in this case it will not be. It is a valid probability if one does not consider it as a single categorical prediction, but rather as independent predictions.

$$P(w^j \mid w^i) = \sigma(V\,C_{:,w^i})_{w^j} \tag{3.49}$$

$$\text{i.e.}\quad P(w^j \mid w^i) = \sigma(V_{w^j,:}C_{:,w^i}) \tag{3.50}$$

Lets start from the cross-entropy loss. In training word w^j does occur near w^i, we know this because they are a training pair presented from the training dataset $\mathcal{X}$. Therefore, since it occurs, we could make a loss function based on minimising the negative log-likelihood of all observations.

$$Loss = \sum_{\forall (w^i, w^j) \in \mathcal{X}} -\log P(w^j \mid w^i) \tag{3.51}$$

This is the cross-entropy loss, excluding the scaling factor for how often it occurs.

Loss Function
Readers may want to reread Sect. 3.4.1.3 to brush up on how we use the training dataset as a ground truth probability estimate implicitly when using cross-entropy loss. When doing so one should remember that that the conditioning term, z, for skip-grams is the co-located words as there is no hidden layer.

However, we are not using softmax in the model output, which means that there is no trade off for increasing (for example) $P(w^1 \mid w^i)$ versus $P(w^2 \mid w^i)$. This thus admits the trivially optimal solution $\forall w^j \in \mathbb{V} \ P(w^j \mid w^i) = 1$. This is obviously wrong – even beyond not being a proper distribution – some words are more commonly co-occurring than others.

So from this we can improve the statement. What is desired from the loss function is to reward models that predict the probability of words that *do* co-occur as being higher, than the probability of words that *do not*. We know that w^j does occur near w^i as it is in the training set. Now, let us select via some arbitrary means a w^k that does not – a negative sample. We want the loss function to be such that $P(w^k \mid w^i) < P(w^j \mid w^i)$. So for this single term in the loss we would have:

$$loss(w^j, w^i) = \log P(w^k \mid w^i) - \log P(w^j \mid w^i) \tag{3.52}$$

Most words do not co-occur
Some simple reasoning can account for this as a reasonable consequence of Zipf's law (Zipf 1949) and a prior of the principle of indifference, but there is a further depth to it as explained by Ha et al. (2009).

The question is then: how is the negative sample w^k to be found? One option would be to deterministically search the corpus for these negative samples, making sure to never select words that actually do co-occur. However that would require enumerating the entire corpus. We can instead just pick them randomly, we can sample from the unigram distribution. As statistically, in any given corpus most words do not co-occur, a randomly selected word in all likelihood will not be one that truly does co-occur – and if it is, then that small mistake will vanish as noise in the training, overcome by all the correct truly negative samples.

Is Eq. (3.53) a function?
No, at the point at which the Loss started including randomly selected samples, it ceased to be a function in the usual mathematical sense. It is still a function in the common computer programming sense though – it is just not deterministic.

At this point, we can question, why limit ourselves to one negative sample? We could take many, and do several at a time, and get more confidence that $P(w^j \mid w^i)$

is indeed greater than other (non-existent) co-occurrence probabilities. This gives the improved loss function of

$$
loss(w^j, w^i) = \left(\sum_{\forall w^k \in \text{samples}(D^{1g})} \log P(w^k \mid w^i) \right) - \log P(w^j \mid w^i) \qquad (3.53)
$$

where D^{1g} stands for the unigram distribution of the vocabulary and samples(D^{1g}) is a function that returns some number of samples from it.

Consider, though is this fair to the samples? We are taking them as representatives of all words that do not co-occur. Should a word that is unlikely to occur at all, *but was unlucky enough to be sampled*, contribute the same to the loss as a word that was very likely to occur? More reasonable is that the loss contribution should be in proportion to how likely the samples were to occur. Otherwise it will add unexpected changes and result in noisy training. Adding a weighting based on the unigram probability $(P^{1g}(w^k))$ gives:

$$
loss(w^j, w^i) = \left(\sum_{\forall w^k \in \text{samples}(D^{1g})} P^{1g}(w^k) \log P(w^k \mid w^i) \right) - \log P(w^j \mid w^i) \qquad (3.54)
$$

The expected value is defined by

$$
\mathbb{E}_{X \sim D}[f(x)] = \sum_{\forall x \text{ values for } X} P^d f(x) \qquad (3.55)
$$

In an abuse of notation, we apply this to the samples, as a sample expected value and write:

$$
\sum_{k=1}^{k=n} \mathbb{E}_{w^k \sim D^{1g}}[\log P(w^k \mid w^i)] \qquad (3.56)
$$

to be the sum of the n samples expected values. This notation (abuse) is as used in Mikolov et al. (2013b). It gives the form:

$$
loss(w^j, w^i) = \left(\sum_{k=1}^{k=n} \mathbb{E}_{w^k \sim D^{1g}}[\log P(w^k \mid w^i)]) \right) - \log P(w^j \mid w^i) \qquad (3.57)
$$

Consider that the choice of unigram distribution for the negative samples is not the only choice. For example, we might wish to increase the relative occurrence of rare words in the negative samples, to help them fit better from limited training data. This is commonly done via subsampling in the positive samples (i.e. the training cases)). So we replace D^{1g} with D^{ns} being the distribution of negative samples from the vocabulary, to be specified as a hyper-parameter of training.

Mikolov et al. (2013b) uses a distribution such that

$$P^{D^{ns}}(w^k) = \frac{P^{D^{lg}}(w^k)^{\frac{2}{3}}}{\sum_{\forall w^o \in \mathbb{V}} P^{D^{lg}}(w^o)^{\frac{2}{3}}} \tag{3.58}$$

which they find to give better performance than the unigram or uniform distributions. Using this, and substituting in the sigmoid for the probabilities, this becomes:

$$loss(w^j, w^i) = \left(\sum_{\substack{k=1 \\ w^k \sim D^{ns}}}^{k=n} \mathbb{E}[\log \sigma(V_{w^k,:}.C_{:,w^i})] \right) - \log \sigma(V_{w^j,:}.C_{:,w^i}) \tag{3.59}$$

By adding a constant we do not change the optimal value. If we add the constant $-K$, we can subtract 1 in each sample term.

$$loss(w^j, w^i) = \left(\sum_{\substack{k=1 \\ w^k \sim D^{ns}}}^{k=n} \mathbb{E}[-1 + \log \sigma(V_{w^k,:}.C_{:,w^i})] \right) - \log \sigma(V_{w^j,:}.C_{:,w^i}) \tag{3.60}$$

Finally we make use of the identity $1 - \sigma(\tilde{z}) = \sigma(-\tilde{z})$ giving:

$$loss(w^j, w^i) = -\log \sigma(V_{w^j,:}.C_{:,w^i}) - \sum_{\substack{k=1 \\ w^k \sim D^{ns}}}^{k=n} \mathbb{E}[\log \sigma(-V_{w^k,:}.C_{:,w^i})] \tag{3.61}$$

Calculating the total loss over the training set $\mathcal{X}$:

$$Loss = -\sum_{\forall (w^i, w^j) \in \mathcal{X}} \left(\log \sigma(V_{w^j,:}.C_{:,w^i}) + \sum_{\substack{k=1 \\ w^k \sim D^{ns}}}^{k=n} \mathbb{E}[\log \sigma(-V_{w^k,:}.C_{:,w^i})] \right) \tag{3.62}$$

This is the negative sampling loss function used in Mikolov et al. (2013b). Perhaps the most confusing part of this is the notation. Without the abuses around expected value, this is written:

$$Loss = -\sum_{\forall (w^i, w^j) \in \mathcal{X}} \left(\log \sigma(V_{w^j,:}.C_{:,w^i}) + \sum_{\forall w^k \in samples(D^{ns})} P^{D^{ns}}(w^k) \log \sigma(-V_{w^k,:}.C_{:,w^i}) \right) \tag{3.63}$$

3.5   Natural Language Applications – Beyond Language Modeling

While statistical language models are useful, they are of-course in no way the be-all and end-all of natural language processing. Simultaneously with the developments around representations for the language modelling tasks, work was being done on solving other NLP problems using similar techniques (Collobert and Jason 2008).

3.5.1   Using Word Embeddings as Features

Pretrained Word-Embeddings

Pretrained Word Embeddings are available for most models discussed here. They are trained on a lot more data than most people have reasonable access to. It can be useful to substitute word embeddings as a representation in most systems, or to use them as initial value for neural network systems that will learn them as they train the system as a whole. There are many many online pretrained word embeddings. One of the more recent and comprehensive set is that of Bojanowski et al. (2016) (based on a skip-gram extension), https://fasttext.cc/docs/en/pretrained-vectors.html They provide embeddings for 294 languages, trained on Wikipedia based on the work of which is an extension to skip-grams.

Turian et al. (2010) discuss what is now perhaps the most important use of word embeddings. The use of the embeddings as features, in unrelated feature driven models. One can find word embeddings using any of the methods discussed above. These embeddings can be then used as features instead of, for example bag of words or hand-crafted feature sets. Turian et al. (2010) found improvements on the state of the art for chunking and Named Entity Recognition (NER), using the word embedding methods of that time. Since then, these results have been superseded again using newer methods.

3.6   Aligning Vector Spaces Across Languages

Given two vocabulary vector spaces, for example one for German and one for English, a natural and common question is if they can be aligned such that one has a single vector space for both. Using canonical correlation analysis (CCA) one can do exactly that. There also exists generalised CCA for any number of vector spaces (Fu et al. 2016), as well as kernel CCA for a non-linear alignment.

The inputs to CCA, are two sets of vectors, normally expressed as matrices. We will call these: $C \in \mathbb{R}^{n^C \times m^C}$ and $V \in \mathbb{R}^{n^V \times m^V}$. They are both sets of vector representations, not necessarily of the same dimensionality. They could be the output of any of the embedding models discussed earlier, or even a sparse (non-embedding) representations such as the point-wise mutual information of the co-occurrence counts. The other input is series pairs of elements from within those those sets that are to be aligned. We will call the elements from that series of pairs from the original sets $C^\star$ and $V^\star$ respectively. $C^\star$ and $V^\star$ are subsets of the original sets, with the same number of representations. In the example of applying this to translation, if each vector was a word embedding: $C^\star$ and $V^\star$ would contains only words with a single known best translation, and this does not have to be the whole vocabulary of either language.

By performing CCA one solves to find a series of vectors (also expressed as a matrix), $S = [\tilde{s}^1 \dots \tilde{s}^d]$ and $T = [\tilde{t}^1 \dots \tilde{t}^d]$, such that the correlation between $C^\star \tilde{s}^i$ and $V^\star \tilde{t}^i$ is maximised, with the constraint that for all $j < i$ that $C^\star \tilde{s}^i$ is uncorrelated with $C^\star \tilde{s}^j$ and that $V^\star \tilde{t}^i$ is uncorrelated with $V^\star \tilde{t}^j$. This is very similar to principal component analysis (PCA), and like PCA the number of components to use (d) is a variable which can be decreased to achieve dimensionality reduction. When complete, taking S and T as matrices gives projection matrices which project C and V to a space where aligned elements are as correlated as possible. The new common vector space embeddings are given by: CS and VT. Even for sparse inputs the outputs will be dense embeddings.

Faruqui and Chris (2014) investigated this primarily as a means to use additional data to improve performance on monolingual tasks. In this, they found a small and inconsistent improvement. However, we suggest it is much more interesting as a multi-lingual tool. It allows similarity measures to be made between words of different languages. Gujral et al. (2016) use this as part of a hybrid system to translate out of vocabulary words. Klein et al. (2015) use it to link word-embeddings with image embeddings.

Dhillon et al. (2011) investigated using this to create word-embeddings. We noted in Eq. (3.16), that skip-gram maximise the similarity of the output and input embeddings according to the dot-product. CCA also maximises similarity (according the correlation), between the vectors from one set, and the vectors for another. As such given representations for two words from the same context, initialised randomly, CCA could be used repeatedly to optimise towards good word embedding capturing shared meaning from contexts. This principle was used by Dhillon et al. (2011), though their final process more complex than described here. It is perhaps one of the more unusual ways to create word embeddings as compared to any of the methods discussed earlier.

Aligning embeddings using linear algebra after they are fully trained is not the only means to end up with a common vector space. One can also directly train embeddings on multiple languages concurrently as was done in Shi et al. (2015), amongst others. Similarly, on the sentence embedding side (Zou et al. 2013), and (Socher et al. 2014) train embeddings from different languages and modalities (respectively) directly to be near to their partners (these are discussed in Chap. 5). A survey paper on such methods was recently published by Ruder (2017).

References

Bengio, Yoshua, Réjean Ducharme, Pascal Vincent, and Christian Janvin. 2003. A neural probabilistic language model. *The Journal of Machine Learning Research*, 137–186.

Bezanson, Jeff, Alan Edelman, Stefan Karpinski, and Viral B. Shah. 2014. Julia: A fresh approach to numerical computing. arXiv:1411.1607 [cs.MS]

Blei, David M., Andrew Y. Ng, and Michael I. Jordan. 2003. Latent Dirichlet allocation. *The Journal of Machine Learning Research* 3: 993–1022.

Bojanowski, Piotr, Edouard Grave, Armand Joulin, and Tomas Mikolov. 2016. Enriching word vectors with subword information. arXiv:1607.04606.

Bolukbasi, Tolga, Kai-Wei Chang, James Y. Zou, Venkatesh Saligrama, and Adam T. Kalai. 2016. Man is to computer programmer as woman is to homemaker? Debiasing word embeddings. In *Advances in Neural Information Processing Systems*, 4349–4357.

Brown, Peter F., Peter V. Desouza, Robert L. Mercer, Vincent J. Della, and Pietra, and Jenifer C. Lai. 1992. Class-based n-gram models of natural language. *Computational Linguistics* 18 (4): 467–479.

Caliskan, Aylin, Joanna J. Bryson, and Arvind Narayanan. 2017. Semantics derived automatically from language corpora contain human-like biases. *Science* 356 (6334), 183–186. ISSN: 0036-8075, https://doi.org/10.1126/science.aal4230, http://science.sciencemag.org/content/356/6334/183.full.pdf.

Collobert, Ronan and Jason Weston. 2008. A unified architecture for natural language processing: Deep neural networks with multitask learning. In *Proceedings of the 25th international conference on Machine learning*, 160–167. ACM.

Cotterell, Ryan, Adam Poliak, Benjamin Van Durme, and Jason Eisner. 2017. Explaining and generalizing skip-gram through exponential family principal component analysis. In *EACL 2017*, 175.

Dhillon, Paramveer, Dean P. Foster, and Lyle H. Ungar. 2011. Multi-view learning of word embeddings via CCA. In *Advances in Neural Information Processing Systems*, 199–207.

Dumais, Susan T., George W. Furnas, Thomas K. Landauer, Scott Deerwester, and Richard Harshman. 1988. Using latent semantic analysis to improve access to textual information. In *Proceedings of the SIGCHI conference on human factors in computing systems*, 281–285. ACM.

Faruqui, Manaal and Chris Dyer. 2014. Improving vector space word representations using multi lingual correlation. In *Association for computational linguistics*.

Fu, X., K. Huang, E. E. Papalexakis, H. A. Song, P. P. Talukdar, N. D. Sidiropoulos, C. Faloutsos, and T. Mitchell. 2016. Efficient and distributed algorithms for large-scale generalized canonical correlations analysis. In *2016 IEEE 16th international conference on data mining (ICDM)*, 871–876. https://doi.org/10.1109/ICDM.2016.0105.

Gladkova, Anna, Aleksandr Drozd, and Satoshi Matsuoka. 2016. Analogy-based detection of morphological and semantic relations with word embeddings: What works and what doesn't. In *SRW@ HLT-NAACL*, 8–15.

Goodman, Joshua. 2001. A bit of progress in language modeling. In *CoRR*. arXiv:cs.CL/0108005.

Gujral, Biman, Huda Khayrallah, and Philipp Koehn. 2016. Translation of unknown words in low resource languages. In *Proceedings of the conference of the association for machine translation in the Americas (AMTA)*.

Gutmann, Michael U. and Aapo Hyvärinen. 2012. Noise-contrastive estimation of unnormalized statistical models, with applications to natural image statistics. *Journal of Machine Learning Research* 13: 307–361.

Ha, Le Quan, Philip Hanna, Ji Ming, and F. Jack Smith. 2009. Extending Zipf's law to n-grams for large corpora. *Artificial Intelligence Review* 32 (1): 101–113.

Huffman, DavidA. 1952. A method for the construction of minimum-redundancy codes. *Proceedings of the IRE* 40 (9): 1098–1101.

Jozefowicz, Rafal, Wojciech Zaremba, and Ilya Sutskever. 2015. An empirical exploration of recurrent network architectures. In *Proceedings of the 32nd international conference on machine learning (ICML-15)*, 2342–2350.

Katz, Slava M. 1987. Estimation of probabilities from sparse data for the language model component of a speech recognizer. *IEEE Transactions on Acoustics, Speech and Signal Processing* 35 (3): 400–401.

Klein, Benjamin, Guy Lev, Gil Sadeh, and Lior Wolf. 2015. Associating neural word embeddings with deep image representations using fisher vectors. In *Proceedings of the IEEE conference on computer vision and pattern recognition*, 4437–4446.

Kneser, Reinhard and Hermann Ney. 1995. Improved backing-off for M-gram language modeling. In *1995 international conference on acoustics, speech, and signal processing, 1995. ICASSP-95*, vol. 1, 181–184. IEEE

Kolmogorov, Vladimir. 2009. Blossom V: A new implementation of a minimum cost perfect matching algorithm. *Mathematical Programming Computation* 1 (1): 43–67.

Landgraf, Andrew J. and Jeremy Bellay. 2017. Word2vec skip-gram with negative sampling is a weighted logistic PCA. In *CoRR*. arXiv:1705.09755.

Levy, Omer and Yoav Goldberg. 2014. Neural word embedding as implicit matrix factorization. In *Advances in neural information processing systems*, 2177–2185.

Levy, Omer, Yoav Goldberg, and Ido Dagan. 2015. Improving distributional similarity with lessons learned from word embeddings. *Transactions of the Association for Computational Linguistics* 3: 211–225. ISSN: 2307-387X.

Li, Yitan, Linli Xu, Fei Tian, Liang Jiang, Xiaowei Zhong, and Enhong Chen. 2015. Word embedding revisited: A new representation learning and explicit matrix factorization perspective. In *IJCAI*, 3650–3656.

Lin, Yuri, Jean-Baptiste Michel, Erez Lieberman Aiden, Jon Orwant, Will Brockman, and Slav Petrov. 2012. Syntactic annotations for the google books ngram corpus. In *Proceedings of the ACL 2012 system demonstrations*, 169–174. Association for Computational Linguistics.

Maaten, Laurens van der and Geoffrey Hinton. 2008. Visualizing data using t-SNE. *Journal of Machine Learning Research* 9: 2579–2605.

Mikolov, Tomas, Martin Karafiát, Lukas Burget, Jan Cernocký, and Sanjeev Khudanpur. 2010. Recurrent neural network based language model. *Interspeech* 2: 3.

Mikolov, Tomas, Kai Chen, Greg Corrado, and Jeffrey Dean. 2013a. Efficient estimation of word representations in vector space. arXiv:1301.3781.

Mikolov, Tomas, Ilya Sutskever, Kai Chen, Greg S Corrado, and Jeff Dean. 2013b. Distributed representations of words and phrases and their compositionality. In *Advances in neural information processing systems*, 3111–3119.

Mikolov, Tomas, Wen-tau Yih, and Geoffrey Zweig. 2013. Linguistic regularities in continuous space word representations. In *HLT-NAACL*, 746–751.

Morin, Frederic and Yoshua Bengio. 2005. Hierarchical probabilistic neural network language model. In *Proceedings of the international workshop on artificial intelligence and statistics* (Citeseer), 246–252.

Pennington, Jeffrey, Richard Socher, and Christopher D. Manning. 2014. GloVe: Global vectors for word representation. In *Proceedings of the 2014 conference on empirical methods in natural language processing (EMNLP 2014)*, 1532–1543.

Rosenfeld, Ronald. 2000. Two decades of statistical language modeling: Where do we go from here? *Proceedings of the IEEE* 88 (8): 1270–1278. https://doi.org/10.1109/5.880083.

Ruder, Sebastian. 2017. A survey of cross-lingual embedding models. In *CoRR*. arXiv:1706.04902.

Schwenk, Holger. 2004. Efficient training of large neural networks for language modeling. In *2004. Proceedings. 2004 IEEE international joint conference on neural networks*, vol. 4, 3059–3064. IEEE.

Shi, Tianze, Zhiyuan Liu, Yang Liu, and Maosong Sun. 2015. Learning cross-lingual word embeddings via matrix co-factorization. In *ACL* (2), 567–572.

Socher, Richard, Andrej Karpathy, Quoc V. Le, Christopher D. Manning, and Y.Ng Andrew. 2014. Grounded compositional semantics for finding and describing images with sentences. *Transactions of the Association for Computational Linguistics* 2: 207–218.

Sundermeyer, Martin, Ralf Schlüter, and Hermann Ney. 2012. LSTM neural networks for language modeling. In *Thirteenth annual conference of the international speech communication association*.

Tomas, Mikolov. 2012. Statistical language models based on neural networks. Ph.D. thesis, Brno University of Technology.

Turian, Joseph, Lev Ratinov, and Yoshua Bengio. 2010. Word representations: A simple and general method for semi-supervised learning. In *Proceedings of the 48th annual meeting of the association for computational linguistics*, 384–394. Association for Computational Linguistics.

Yang, Zhixuan, Chong Ruan, Caihua Li, and Junfeng Hu. 2016. Optimize hierarchical softmax with word similarity knowledge. In *17th international conference on intelligent text processing and computational linguistics (CICLing)*.

Zipf, G.K. 1949. *Human behavior and the principle of least effort: An introduction to human ecology*. Cambridge: Addison-Wesley Press.

Zou, Will Y., Richard Socher, Daniel M. Cer, and Christopher D. Manning. 2013. Bilingual word embeddings for phrase-based machine translation. In *EMNLP*, 1393–1398.

Word Sense Representations

4

1a. In a literal, exact, or actual sense; not figuratively, allegorically, etc
1b. Used to indicate that the following word or phrase must be taken in its literal sense, usually to add emphasis
1c. colloq. Used to indicate that some (frequently conventional) metaphorical or hyperbolical expression is to be taken in the strongest admissible sense: 'virtually, as good as'; (also) 'completely, utterly, absolutely' ...
2a With reference to a version of something, as a transcription, translation, etc.: in the very words, word for word
2b. In extended use. With exact fidelity of representation; faithfully
3a. With or by the letters (of a word). Obs. rare
3b. In or with regard to letters or literature. Obs. rare

The seven senses of literally, Oxford English
Dictionary, 3rd ed., 2011

Abstract

In this chapter, techniques for representing the multiple meanings of a single word are discussed. This is a growing area, and is particularly important in languages where polysemous and homonymous words are common. This includes English, but it is even more prevalent in Mandarin for example. The techniques discussed can broadly be classified as lexical word sense representation, and as word sense induction. The inductive techniques can be sub-classified as clustering-based or as prediction-based.

4.1 Word Senses

Words have multiple meanings. A single representation for a word cannot truly describe the correct meaning in all contexts. It may have some features that are applicable to some uses but not to others, it may be an average of all features for all uses, or it may only represent the most common sense. For most word-embeddings it will be an unclear combination of all of the above. Word sense embeddings attempt to find representations not of words, but of particular senses of words.

Polysemous/Homonymous
A word with multiple meanings i.e. senses. For NLP representational purposes polysemous and homonymous are synonymous.

The standard way to assign word senses is via some lexicographical resource, such as a dictionary, or a thesaurus. There is not a canonical list of word senses that are consistently defined in English. Every dictionary is unique, with different definitions and numbers of word senses. The most commonly used lexicographical resource is WordNet (Miller et al. 1995), and the multi-lingual BabelNet (Navigli and Simone 2010). The relationship between the terminology used in word sense problems is shown in Fig. 4.1.

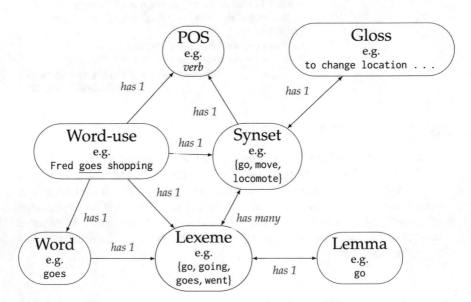

Fig. 4.1 The relationship between terms used to discuss various word sense problems. The lemma is used as the representation for the lexeme, for WordNet's purposes when indexing. For many tasks each the word-use is pre-tagged with its lemma and POS tag, as these can be found with high reliability using standard tools. Note that the arrows in this diagram are *directional*. That is to say, for example, each Synset *has 1* POS, but each POS *has many* Synsets

Part of Speech/POS
The syntactic category a word belongs to. Different POS tags come from different tag sets. Can be simple as the WordNet tag set: noun, adjective, verb, etc. or complex as in the Brown tag set: VBG-*verb, gerund/present participle*, NN-*noun, singular or mass*.

Word-use
An occurrence of a word in a text, such as a training corpus. Each word will have multiple uses in a text. Each word-use will only have one particular meaning and will thus belong to one synset.

Lemma
The base form of the word as defined by a lexicographical resource. It is normally closely related to (often identical to) the *stem* which is the word's root form with all morphological inflections (e.g. tenses) removed.

Lexeme
The set of words that share a common lemma: go, going, goes, and went all belong to the lexeme headed by the lemma go

Synset
A synset is a set of synonymous words: that is words that have the same meaning. In lexicographic terms the synset is the core unit of meaning. Identifying the synset of a word-use is the the same as identifying the word sense. Every word sense corresponds to one synset.

Gloss
A gloss is the dictionary entry for a word sense, it normally includes both the definition and an example of use. In WordNet each synset shares a common gloss.

4.1.1 Word Sense Disambiguation

Word sense disambiguation is one of the hardest problems in NLP. Very few systems significantly out perform the baseline, i.e. the most frequent sense (MFS) technique.

Progress on the problem is made difficult by several factors.

The sense is hard to identify from the context. Determining the sense may require very long range information: for example the information on context may not even be in the same sentence. It may require knowing the domain of the text, because word sense uses vary between domains. Such information is external to the text itself. It may in-fact be intentionally unclear, with multiple correct interpretations, as in a pun. It maybe unintentionally unknowable, due to a poor writing style, such that it would confuse any human reader. These difficulties are compounded by the limited amount of data available.

There is only a relatively small amount of labelled data for word sense problems. It is the general virtue of machine learning that given enough data, almost any input-output mapping problem (i.e. function approximation) can be solved. Such an amount of word sense annotated data is not available. This is in contrast to finding unsupervised word embeddings, which can be trained on any text that has ever been written. The lack of very large scale training corpora renders fully supervised methods difficult. It also results in small sized testing corpora; which leads to systems that may appear to perform well (on those small test corpora), but do not generalise to real world uses. In addition, the lack of human agreement on the correct sense, resulting in weak ground truth, further makes creating new resources harder. This limited amount of data compounds the problem's inherent difficulties.

It can also be said that word senses are highly artificial and do not adequately represent meaning. However, WSD is required to interface with lexicographical resources, such as translation dictionaries (e.g. BabelNet), ontologies (e.g. OpenCyc), and other datasets (e.g. ImageNet Deng et al. 2009).

It may be interesting to note, that the number of meanings that a word has is approximately inversely proportional related to its frequency of use rank (Zipf 1945). That is to say the most common words have far more meanings than rarer words. It is related to (and compounds with) the more well-known Zipf's Law on word use (Zipf 1949), and can similarly be explained-based on Zipf's core premise of the principle of least effort. This aligns well with our notion that precise (e.g. technical) words exist but are used only infrequently – since they are only explaining a single situation. This also means that by most word-uses are potentially very ambiguous.

The most commonly used word sense (for a given word) is also overwhelmingly more frequent than its less common brethren – word sense usage also being roughly Zipfian distributed (Kilgarriff 2004). For this reason the Most Frequent Sense (MFS) is a surprisingly hard baseline to beat in any WSD task.

Lemmatization
Lemmatization is the method of converting a word into its lemma. Due to the similarly of the lemma to the stem, this in essence means removing the tense and plurality information (stemming), with some additional special-cases. Word-Net is indexed by lemmas, and comes with a lemmatizer called morphy allowing any word to be looked up by lemmatizing it to it's lemma

Unlemmatization

Given lemma (as one can extract from WordNet) and a full POS tag (such as a Brown-style tag) for a word, it is possible to undo the lemmatization with a high degree of reliability using relatively simple rules (again due to the similarity of the lemma to the stem). The POS tag encodes the key inflectional features that are lost. Patten.en (De Smedt and Walter 2012) is a python library encoding such rules (pluralisation, verb conjugation, etc.); though combining them with the POS tag to drive them is a task left for the reader. This can be used to find substitute words using WordNet's features, for finding synonyms, antonyms and other lemmas from lexically related categories.

Semantic Syllepsis

(Also known as pathological sentences that kill almost all WSD systems.) Consider the sentence: John used to work for the *newspaper* that you are carrying.. In this sentence the word-use newspaper simultaneously have two different meanings: it is both the company, and the object. This violates our earlier statement that every word-use belongs to exactly one synset. WSD systems are unable to handle these sentences as they attempt to assign a single sense to each word-use. Most word sense induction systems cannot do much better: at best a new sense could be allocated for the joint use, which does not correspond to the linguistic notion of the word having two senses for different parts of the sentence. Most works on word sense disambiguation outright ignore these sentences, or consider them to be ungrammatical, or incorrect. However, they are readily understood and used without thought by most native speakers. These constructions are also known as *zeugma*, although zeugma is itself a highly polysemous word, so its usage varies.

4.1.1.1 Most Frequent Sense

Given a sense annotated corpus, it is easy to count how often each sense of a word occurs. Due to the over-whelming frequency of the most frequent sense, it is unlikely for even a small training corpus to have the most frequent sense differing from the use in the language as a whole.

The Most Frequent Sense (MFS) method of word sense disambiguation is defined by counting the frequency of a particular word sense for a particular POS tagged word. For the ith word use being the word w^i, having some sense s^j then without any further context the probability of that sense being the correct sense is $P(s^j \mid w^i)$. One can use the part of speech tag p_i (for the ith word use) as an additional condition, and thus find $P(s^j \mid w^i, p_i)$. WordNet encodes this information for each lemma-synset pair (i.e. each word sense) using the SemCor corpus counts. This is also used for sense ordering, which is why most frequent sense is sometimes called first sense.

This is a readily available and practical method for getting a baseline probability of each sense. Most frequent sense can be applied for word sense disambiguation using this frequency-based probability estimate: $\text{argmax}_{\forall_{s^j}} P(s^j \mid w^i, p_i)$.

In the most recent SemEval WSD task (Moro and Roberto 2015), MFS beat all submitted entries for English, both overall, and on almost all cuts of the data. The results for other languages were not as good, however in other languages the true corpus-derived sense counts were not used.

WordNet is not a strong moral baseline

WordNet, as a resource,-based partly on the work of Princeton undergraduate students in the early 1990s, and on the literature of 1961, is not the kind of resource one might hope for from an AI information perspective. The glosses include a number of biases. These biases are reflective of the language use, but are not necessarily ideal to be encoded into a system. For example: `S: (v) nag, peck, hen-peck (bother persistently with trivial complaints) 'She nags her husband all day long's`. Other dictionaries regularly show up in the News for similar content.

Another problem is the source of the word sense counts. As discussed in the main text, sense counts are important in WSD systems. The counts come from SemCor, a sense annotated subset of the Brown Corpus. The Brown Corpus is a sampling of American texts from 1961. The cultural norms of 1961 were not the norms of today. (For context, note that the US did not pass the Civil Rights act to end segregation until 1964). As such, one should not trust WordNet (or SemCor) to reflect current sense counts, for words which have undergone usage change since 1961.

Furthermore, when creating down stream resources-based on WordNet, one should not use these sense counts to determine how important it is to include a concept. If ImageNet (Deng et al. 2009) for example, had used SemCor counts to determine which synsets of images would be included, then items rarely discussed in 1961 literature, like `wheelchairs`, and `prosthesis` would be excluded. Which would in turn make many image processing systems systematically unhelpful in processing images relating to the disabled. (Do not fear: even the initial release of ImageNet contains hundreds of images of `wheelchairs`, and `prosthesis`) Unintentional biasing of data can have on-going effects on the behaviour of machine learning-based systems far beyond the original conception.

4.2 Word Sense Representation

It is desirable to create a vector representation of a word sense much like in Chap. 3 representations were created for words. We desire to an embedding to represent each word sense, as normally represented by a word-synset pair. This section considers the representations for the lexical word senses as given from a dictionary. We consider a direct method of using a labelled corpus, and an indirect method makes use of simpler sense-embeddings to partially label a corpus before retraining. These methods create representations corresponding to senses from WordNet. Section 4.3 considers the case when the senses are to also be discovered, as well as represented.

4.2.1 Directly Supervised Method

The simple and direct method is to take a dataset that is annotated with word senses, and then treat each sense-word pair as if it were a single word, then apply any of the methods for word representation discussed in this chapter. Iacobacci et al. (2015) use a CBOW language model (Mikolov et al. 2013) to do this. This does, however, run into the aforementioned problem, that there is relatively little training data that has been manually sense annotated. Iacobacci et al. (2015) use a third-party WSD tool, namely BabelFly (Moro et al. 2014), to annotate the corpus with senses. This allows for existing word representation techniques to be applied.

Chen et al. (2014) applies a similar technique, but using a word-embedding-based partial WSD system of their own devising, rather than an external WSD tool.

4.2.2 Word Embedding-Based Disambiguation Method

Chen et al. (2014) uses an almost semi-supervised approach to train sense vectors. They partially disambiguate their training corpus, using initial word sense vectors and WordNet. They then completely replace these original (phase one) sense-vectors, by using the partially disambiguated corpus to train new (phase two) sense-vectors via a skip-gram variant. This process is shown in Fig. 4.2.

The **first phase** of this method is in essence a word-embedding-based WSD system. When assessed as such, they report that it only marginally exceeds the MFS baseline, though that is not at all unusual for WSD algorithms as discussed above.

They assign a sense vector to every word sense in WordNet. This sense vector is the average of word-embeddings of a subset of words in the gloss, as determined using pretrained skip-grams (Mikolov et al. 2013). For the word w with word sense w^{s^i}, a set of candidate words, $cands(w^{s^i})$, is selected from the gloss based on the following set of requirements. First, the word must be a content word: that is a verb, noun, adverb or adjective; secondly, its cosine distance to w must be below some threshold δ; finally, it must not be the word itself. When these requirements are followed $cands(w^{s^i})$ is a set of significant closely related words from the gloss.

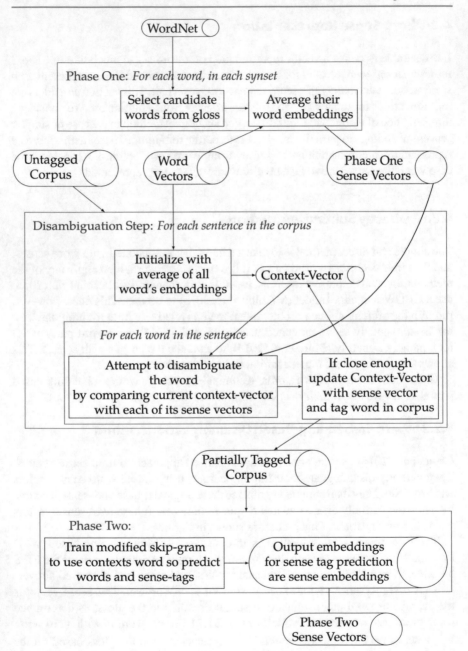

Fig. 4.2 The process used by Chen et al. (2014) to create word sense embeddings

WSD with embeddings
It is beyond the scope of this work to fully discuss WSD systems. However, we will remark that (single sense) word embeddings are a generally useful feature as an input to any NLP ML system. As such they can be used as features in a fully supervised WSD system. The idea of using them in this way is is similar to the LSI enhanced Lesk WSD system of Basile et al. (2014).

Cosine distance
Here we talk of cosine distance, where a smaller distance implies more similar (and 0.0 identical). Contrasting this with the cosine similarity, where higher value implies more similar (and 1.0 identical).
Cosine distance is still not a true metric as $d^{cos}(v, kv) = 0$ for all $k \in \mathbb{R}_+$. Other times you may see cosine similarity, ranging between -1 (most different) and 1 (most similar. Cosine similarity is given by $sim(a, b) = \frac{\tilde{a} \cdot \tilde{b}}{\|\tilde{a}\|_2 \|\tilde{b}\|_2} = \cos(\angle \tilde{a}\tilde{b})$ i.e. the unit-length normalised dot product of the vectors.
Cosine distance is usually defined as $d^{cos}(\tilde{a}, \tilde{b}) = \frac{1 - sim(\tilde{a}, \tilde{b})}{2}$. Ranging between 0 (most similar) and 1 (most different).

The phase one sense vector for w^{s^i} is the mean of the word vectors for all the words in $cands(w^{s^i})$. The effect of this is that we expect that the phase one sense vectors for most words in the same synset will be similar but not necessarily identical. This expectation is not guaranteed however. As an example, consider the use of the word china as a synonym for porcelain: the single sense vector for china will likely be dominated by its more significant use referring to the country, which would cause very few words in the gloss for the porcelain synset to be included in cands. Resulting in the phase one sense vectors for the synonymous senses of porcelain and china actually being very different.

The phase one sense vectors are used to disambiguate the words in their unlabelled training corpus. For each sentence in the corpus, an initial *content vector* is defined by taking the mean of the skip-gram word embedding (not word sense) for all content words in the sentence. For each word in the sentence, each possible sense-embedding is compared to the context vector. If one or more senses vectors are found to be closer than a given threshold, then that word is tagged with the closest of those senses, and the context vector is updated to use the sense-vector instead of the word vector. Words that do not come within the threshold are not tagged, and the context vector is not updated. This is an important part of their algorithm, as it ensures that words without clear senses do not get a sense ascribed to them. This thus avoids any dubious sense tags for the next training step.

In **phase two** of training (Chen et al. 2014) employ the skip-gram word-embedding method, with a variation, to predict the word senses. They train it on the partially disambiguated corpus produced in phase one. The original sense vectors are discarded. Rather than the model being tasked only to predict the surrounding words, it is tasked to predict surrounding words and their sense-tags (where present). In the loss function the prediction of tags and words is weighted equally.

Note that the input of the skip-gram is the just central word, not the pair of central word with sense-tag. In this method, the word sense embeddings are output embeddings; though it would not be unreasonable to reverse it to use input embeddings with sense tags, or even to do both. The option to have input embeddings and output embeddings be from different sets, is reminiscent of Schwenk (2004) for word embeddings.

The phase one sense vectors have not been assessed on their representational quality. It could be assumed that because the results for these were not reported, they were worse than those found in phase two. The phase two sense vectors were not assessed for their capacity to be used for word sense disambiguation. It would be desirable to extend the method of Chen et al. (2014), to use the phase two vectors for WSD. This would allow this method to be used to disambiguate its own training data, thus enabling the method to become self-supervised.

4.3   Word Sense Induction (WSI)

Can we go from induced senses to lexical senses?
A natural question given the existence of many WSI systems, and the existing wealth of lexically indexed resources, is if we can align induced senses to a set of lexically defined senses. Agirre et al. (2006) proposed a method for doing this using a weighted mapping-based on the probabilities found using induced sense WSD on a labelled "mapping" corpus. This has only been used on relatively small datasets with only hundreds of words (SenseEval 3 Mihalcea et al. 2004 and SemEval-2007 Task 02 Agirre and Aitor 2007). Our own investigations in White et al. (2018) with the larger SemEval 2007 Task 7 (Navigli et al. 2007) suggest that it may not scale very well to to real-word WSD tasks. That work proposed an alternative method that worked better, though still not as well as could be hoped. Finding suitable methods to link unsupervised representations, to human defined senses remains a topic worthy of research.

In this section we will discuss methods for finding a word sense without reference to a standard set of senses. Such systems must discover the word senses at the same time as they find their representations. One strong advantage of these methods is that they do not require a labelled dataset. As discussed there are relatively few high-quality word sense labelled datasets. The other key advantage of these systems is that they do not rely on fixed senses determined by a lexicographer. This is particularly useful if the word senses are highly domain specific; or in a language without strong lexicographical resources. This allows the data to inform on what word senses exist.

Why do skip-grams perform so well on SCWS?
SCWS is a corpus designed for evaluating word sense embeddings. Single sense embeddings (e.g. skip-grams) cannot take advantage of the context infor-mation in the SCWS. However, they do often perform comparably to the word sense embeddings. Sometimes even outperforming them. It is unclear if this highlights the difficulty of the task (i.e. that the impact of context is hard to gauge), or it might be due to the (implicit) most frequent sense dominating both the use in the tasks, and the representation in a single sense. Alternatively, it may just be the result of the fine tuning of the more mature single-sense embedding methods (and that with more time and tuning multiple sense methods could do proportionally better).

Most vector word sense induction and representation approaches are evaluated on similarity tests. Such tests include WordSim-353 (Finkelstein et al. 2001) for context-less, or Stanford's Contextual Word Similarities (SCWS) for similarity with context information (Huang et al. 2012). This evaluation is also suitable for evaluating single sense word-embeddings, e.g. skip-grams.

We can divide the WSI systems into context clustering-based approaches, and co-location prediction-based approaches. This division is similar to the separation of co-location matrix factorisation, and co-location prediction-based approaches discussed in Chap. 3. It can be assumed thus that at the core, like for word embeddings, they are fundamentally very similar. One could think of prediction of collocated words as a soft indirect clustering of contexts that can have those words.

4.3.1   Context Clustering-Based Approaches

As the meaning of a word, according to word embedding principles, is determined by the contexts in which it occurs, we expect that different meanings (senses) of the same words should occur in different contexts. If we cluster the contexts that a word occurs in, one would expect to find distinct clusters for each sense of the word. It is on this principle that the context clustering-based approaches function.

4.3.1.1   Offline Clustering

The fundamental method for most clustering-based approaches is as per Schütze (1998). That original work is not a neural word sense embedding, however the approach remains the same. Pantel and Dekang (2002) and Reisinger and Raymond (2010) are also not strictly neural word embedding approaches (being more classical vector representations), however the overall method is also very similar.

> **On context representations**
> These co-location clustering methods require finding a representation for the context (from this a similarity metric is applied, and the clustering is then done). More generally, this can be related to the next chapter: Chap. 5, as any of these methods could be used to derive a vector representation of a context. In most works (including all the works discussed here) comparatively simple representations of the contexts are used. It would be interesting to extend the sentence representation methods, and apply them to this use.

The clustering process is done by considering all word uses, with their contexts. The contexts can be a fixed-sized window of words (as is done with many word-embedding models), the sentence, or defined using some other rule. Given a pair of contexts, some method of measuring their similarity must be defined. In vector representational works, this is ubiquitously done by assigning each context a vector, and then using the cosine similarity between those vectors.

The **first step** in all the offline clustering methods is thus to define the representations of the contexts. Different methods define the context vectors differently:

- Schütze (1998) uses variations of inverse-document-frequency (idf) weighted bags of words, including applying dimensionality reduction to find a dense representation.
- Pantel and Dekang (2002) use the mutual information vectors between words and their contexts.
- Reisinger and Raymond (2010), use td-idf or χ^2 weighted bag of words.
- Huang et al. (2012) uses td-idf weighted averages of (their own) single sense word embeddings for all words in the context.
- Kågebäck et al. (2015) also uses a weighted average of single sense word skip-gram embeddings, with the weighting based on two factors. One based on how close the words were, and the other on how likely the co-occurrence was according to the skip-gram model.

It is interesting to note that idf, td-idf, mutual information, skip-gram co-occurrence probabilities (being a proxy for point-wise mutual information Levy and Yoav 2014), are all closely related measures.

On clustering
Clustering can be defined as a (mixed integer) optimisation task, of assigning points to clusters so as to satisfy some loss function-based around minimising intra-cluster variance while maximising inter-cluster variance (or a similar measure). As this is NP-hard, most clustering methods are approximate. K-means is very popular because of its simplicity, however it easily falls into local minima, and so normally it is run dozens of times (at least) to obtain more optimal results. K-means also has the issue of having to select the number of clusters (k). It should be remembered that there exist many other clustering methods than k-means (and its variants). These other methods use different loss functions, and different strategies to overcome the NP-hard nature of the problem. In particular their mixture model methods, hierarchical methods, spectral methods, and others. We personally favour affinity propagation (Frey and Delbert 2007), though there is provably no ideal clustering algorithm even in the non-heuristic case (Kleinberg 2003). On any clustering task (word sense or otherwise) it is worth investigating several clustering algorithms, and not just settling for k-means (particularly not settling for k-means run once.). A series of interesting and easy reading articles on clustering can be found at: http://alexhwilliams.info/itsneuronalblog/2015/09/11/clustering1/, http://alexhwilliams.info/itsneuronalblog/2015/10/01/clustering2/, and http://alexhwilliams.info/itsneuronalblog/2015/11/18/clustering-is-easy/

The **second step** in off-line clustering is to apply a clustering method to cluster the word-uses. This clustering is done based on the calculated similarity of the context representation where the words are used. Again, different WSI methods use different clustering algorithms.

- Schütze (1998) uses a group average agglomerative clustering method.
- Pantel and Dekang (2002) use a custom hierarchical clustering method.
- Reisinger and Raymond (2010) use mixtures of von-Mises-Fisher distributions.
- Huang et al. (2012) use spherical k-means.
- Kågebäck et al. (2015) use k-means.

The **final step** is to find a vector representation of each cluster. For non-neural embedding methods this step is not always done, as defining a representation is not the goal, though in general it can be derived from most clustering techniques. Schütze (1998) and Kågebäck et al. (2015) use the centroids of their clusters. Huang et al. (2012) use a method of relabelling the word uses with a cluster identifier, then train a (single-sense) word embedding method on cluster identifiers rather than words. This relabelling technique is similar to the method later used by Chen et al. (2014) for learning lexical sense representations, as discussed in Sect. 4.2.2. As each cluster of contexts represents a sense, those cluster embeddings are thus also considered as suitable word sense embeddings.

To summarize, all the methods for inducing word sense embeddings via off-line clustering follow the same process. **First**: represent the contexts of word use, so as to be able to measure their similarity. **Second**: use the context's similarity to cluster them. **Finally**: find a vector representation of each cluster. This cluster representation is the induced sense embedding.

4.3.1.2   Online Clustering

The methods discussed above all use off-line clustering. That is to say the clustering is performed after the embedding is trained. Neelakantan et al. (2015) perform the clustering during training. To do this they use a modified skip-gram-based method. They start with a fixed number of randomly initialised sense vectors for each context. These sense vectors are used as input embeddings for the skip-gram context prediction task, over single sense output embeddings. Each sense also has, linked to it, a context cluster centroid, which is the average of all output embeddings for the contexts that the sense is assigned to. Each time a training instance is presented, the average of the context output embeddings is compared to each sense's context cluster centroid. The context is assigned to the cluster with the closest centroid, updating the centroid value. This can be seen as similar to performing a single k-means update step for each training instance. Optionally, if the closest centroid is further from the context vector than some threshold, a new sense can be created using that context vector as the initial centroid. After the assignment of the context to a cluster, the corresponding sense vector is selected for use as the input vector in the skip-gram context prediction task.

Kågebäck et al. (2015) investigated using their weighting function (as discussed in Sect. 4.3.1.1) with the online clustering used by Neelakantan et al. (2015). They found that this improved the quality of the representations. More generally any such weighting function could be used. This online clustering approach is loosely similar to the co-location prediction-based approaches.

4.3.2   Co-location Prediction-Based Approaches

Probability
One may wish to brush up on basic probability notions for this section. In particular joint, conditional and marginal probabilities definitions; as well as Bayes Theorem and the probability chain-rule which come from those. In brief these are as follows.
Conditional Probability: $P(A \mid B) = \frac{P(A,B)}{P(B)}$

Marginal Probability: $P(A) = \sum_{\forall b} P(A, B = b)$

Bayes Theorem: $P(A \mid B) = \frac{P(B|A)P(A)}{P(B)}$

Probability Chain-rule: $P(A^n, \ldots, A^1) = P(A^n \mid A^{n-1}, \ldots, A^1) P(A^{n-1}, \ldots, A^1)$
e.g. $P(A, B, C) = P(A \mid B, C)P(B \mid C)P(C)$
The latter three rules are consequences of the first.

Rather than clustering the contexts, and using those clusters to determine embeddings for different senses, one could consider the sense as a latent variable in the task used to find word embeddings – normally a language modelling task. The principle is that it is not the word that determines its collocated context words, but rather the word sense. So the word sense can be modelled as a hidden variable, where the word, and the context words are being observed.

Tian et al. (2014) used this to define a skip-gram-based method for word sense embeddings. For input word w^i with senses $\mathcal{S}(w^i)$, the probability of output word w^o occurring near w^i can be given as:

$$P(w^o \mid w^i) = \sum_{\forall s^k \in \mathcal{S}(w^i)} P(w^o \mid s^k, w^i) P(s^k \mid w^i) \tag{4.1}$$

Given that a sense s^k only belongs to one word w^i, we know that kth sense of the ith word only occurs when the ith word occurs. We have that the join probability $P(w^i, s^k) = P(s^k)$.

We can thus rewrite Eq. 4.1 as:

$$P(w^o \mid w^i) = \sum_{\forall s^k \in \mathcal{S}(w^i)} P(w^o \mid s^k) P(s^k \mid w^i) \tag{4.2}$$

A softmax classifier can be used to define $P(w^o \mid s^k)$, just like in normal language modelling. With output embeddings for the words w^o, and input embeddings for the word senses s^k. This softmax can be sped-up using negative sampling or hierarchical softmax. The later was done by Tian et al. (2014).

Equation 4.2 is in the form of a mixture model with a latent variable. Such a class of problems are often solved using the Expectation Maximisation (EM) method. In short, the EM procedure functions by performing two alternating steps. The **E-step** calculates the expected chance of assigning word sense for each training case $(\hat{P}(s^l \mid w^o))$ in the training set $\mathcal{X}$. Where a training case is a pairing of a word use w^i, and context word w^o, with $s^l \in \mathcal{S}(w^i)$, formally we have:

$$\hat{P}(s^l \mid w^o) = \frac{\hat{P}(s^l \mid w^i) P(w^o \mid s^l)}{\sum_{\forall s^k \in \mathcal{S}(w^i)} \hat{P}(w^o \mid s^k) P(s^k \mid w^i)} \tag{4.3}$$

The **M-step** updates the prior likelihood of each sense (that is without context) using the expected assignments from the E-step.

$$\hat{P}(s^l \mid w^i) = \frac{1}{|\mathcal{X}|} \sum_{\forall (w^o, w^i) \in \mathcal{X}} \hat{P}(s^l \mid w^o) \tag{4.4}$$

During this step the likelihood of the $P(w^o \mid w^i)$ can be optimised to maximise the likelihood of the observations. This is done via gradient descent on the neural network parameters of the softmax component: $P(w^o \mid s^k)$. By using this EM optimisation the network can fit values for the embeddings in that softmax component.

A limitation of the method used by Tian et al. (2014), is that the number of each sense must be known in advance. One could attempt to solve this by using, for example, the number of senses assigned by a lexicographical resource (e.g. WordNet). However, situations where such resources are not available or not suitable are one of the main circumstances in which WSI is desirable (for example in work using domain specific terminology, or under-resourced languages). In these cases one could apply a heuristic-based on the distribution of senses-based on the distribution of words (Zipf 1945). An attractive alternative would be to allow senses to be determined-based on how the words are used. If they are used in two different ways, then they should have two different senses. How a word is being used can be determined by the contexts in which it appears.

Bartunov et al. (2015) extend on this work by making the number of senses for each word itself a fit-able parameter of the model. This is a rather Bayesian modelling approach, where one considers the distribution of the prior.

WordNet and BabelNet

As mentioned in the previous sections, WordNet and BabelNet are the predominant lexicons used for word senses. It is not directly relevant to this section, but we have space here to remark upon them. WordNet (Tengi et al. 1998) as a very well established tool has a have a binding in practically every modern programming language suitable for NLP. WordNet.jl (https://github.com/JuliaText/WordNet.jl) is the Julia binding. NLTK (Bird et al. 2009) includes one for Python. BabelNet (Navigli and Simone 2010) is intended to be accessed as an online resource, via a RESTful API. Users receive 1000 free queries per day. Academic users can request an upgrade to 50,000 queries per day, or to download a copy of the database. From personal experience we found those requests to be handled easily and rapidly.

Considering again the form of Eq. 4.2

$$P(w^o \mid w^i) = \sum_{\forall s^k \in \mathcal{S}(w^i)} P(w^o \mid s^k) P(s^k \mid w^i) \tag{4.5}$$

The prior probability of a sense given a word, but no context, is $P(s^k \mid w^i)$. This is Dirichlet distributed. This comes from the definition of the Dirichlet distribution as the the the prior probability of any categorical classification task. When considering that the sense my be one from an unlimited collection of possible senses, then that prior becomes a Dirichlet process.

In essence, this prior over a potentially unlimited number of possible senses becomes another parameter of the model (along with the input sense embeddings and output word embeddings). The fitting of the parameters of such a model is beyond

the scope of this book; it is not entirely dissimilar to the fitting via expectation maximisation incorporating gradient descent used by Tian et al. (2014). The final output
of Bartunov et al. (2015) is as desired: a set of induced sense embeddings, and a
language model that is able to predict how likely a word is to occur near that word
sense ($P(w^o \mid s^k)$).

By application of Bayes' theorem, the sense language model can be inverted to
take a word's context, and predict the probability of each word sense.

Independence Assumption
Technically, Eq. 4.6 does not require the independence of the probabilities of
the context words. Rather it only requires that the context words be condition-
ally independent on the word in question w^i. Nevertheless, even the conditional
independence assumption is incorrect, except for a theoretical perfect embedding capturing perfect information. The conditional independence assumption
remains useful as an approximation.

$$P(s^l \mid w^o) = \frac{P(w^o \mid s^l)P(s^l \mid w^i)}{\sum_{\forall s^k \in \mathcal{S}(w^i)} P(w^o \mid s^k)P(s^k \mid w^i)} \tag{4.6}$$

with the common (but technically incorrect) assumption that all words in the context
are independent.

Given a context window: $\mathcal{W}^i = \left(w^{i-\frac{n}{2}}, \ldots, w^{i-1}, w^{i+1}, \ldots, w^{i+\frac{n}{2}} \right)$, we have:

$$P(s^l \mid \mathcal{W}^i) = \frac{\prod_{\forall w^j \in \mathcal{W}^i} P(w^j \mid s^l)P(s^l \mid w^i)}{\sum_{s^k \in \mathcal{S}(w^i)} \prod_{\forall w^j \in \mathcal{W}^i} P(w^j \mid s^k)P(s^k \mid w^i)} \tag{4.7}$$

4.4   Conclusion

Word sense representations allow the representations of the senses of words when one
word has multiple meanings. This increases the expressiveness of the representation.
These representations can in general be applied anywhere word embeddings can.
They are particularly useful for translation, and in languages with large numbers of
homonyms.

The word representation discussions in this chapter naturally lead to the next
section on phrase representation. Rather than a single word having many meanings,
the next chapter will discuss how a single meaning may take multiple words to
express. In such longer structure's representations, the sense embeddings discussed
here are often unnecessary, as the ambiguity may be resolved by the longer structure.
Indeed, the methods discussed in this chapter have relied on that fact to distinguish
the senses using the contexts.

Finding the nearest neighbours (Nearest Neighbour Trees)

A common evaluation task with any representation is to find its nearest neighbours. The naïve solution is to check the distance to all points. For n points this is $O(n)$ operations. For word embeddings n is the size of the vocabulary, perhaps 100,000 words. Performing 100,000 operations per check, is not entirely unreasonable on modern computers (even when the operations are on 300 dimensional representations). However, for word sense embeddings, which have many senses per word in the vocabulary, this means many more points to check. 30 senses per word is not unusual for fine-grained word sense induction. Having a total $n = 3{,}000{,}000$ representations to check causes a noticeable delay. To solve this we can use data structures designed for fast nearest neighbour querying. A k-d tree takes at worst $O(n \log_2(n))$ time to construct. Once constructed on average it takes $O(\log(n))$ to find the nearest neighbour to any point. This makes checking the nearest neighbour nearly instantaneous for even the largest vocabularies.

References

Agirre, Eneko and Aitor Soroa. 2007. Semeval-2007 task 02: Evaluating word sense induction and discrimination systems. In *Proceedings of the 4th international workshop on semantic evaluations. SemEval '07* , 7–12. Prague, Czech Republic: Association for Computational Linguistics.

Agirre, Eneko, David Martínez, Oier López De Lacalle, and Aitor Soroa. 2006. Evaluating and optimizing the parameters of an unsupervised graph-based WSD algorithm. In *Proceedings of the first workshop on graph based methods for natural language processing*, 89–96. Association for Computational Linguistics.

Bartunov, Sergey, Dmitry Kondrashkin, Anton Osokin, and Dmitry P. Vetrov. 2015. Breaking sticks and ambiguities with adaptive skip-gram. In *CoRR*. arXiv:1502.07257.

Basile, Pierpaolo, Annalina Caputo, and Giovanni Semeraro. 2014. An Enhanced lesk word sense disambiguation algorithm through a distributional semantic model. In *Proceedings of COLING 2014, the 25th international conference on computational linguistics: Technical papers. Dublin*, 1591–1600. Ireland: Dublin City University and Association for Computational Linguistics.

Bird, Steven, Ewan Klein, and Edward Loper. 2009. *Natural language processing with Python*. O'Reilly Media, Inc.

Chen, Xinxiong, Zhiyuan Liu, and Maosong Sun. 2014. A unified model for word sense representation and disambiguation. In *EMNLP* (Citeseer), 1025–1035.

De Smedt, Tom and Walter Daelemans. 2012. Pattern for python. *The Journal of Machine Learning Research* 13 (1): 2063–2067.

Deng, J., W. Dong, R. Socher, L.-J. Li, K. Li, and L. Fei-Fei. 2009. ImageNet: A large-scale hierarchical image database. In *CVPR09*.

Finkelstein, Lev, Evgeniy Gabrilovich, Yossi Matias, Ehud Rivlin, Zach Solan, Gadi Wolfman, and Eytan Ruppin. 2001. Placing search in context: The concept revisited. In *Proceedings of the 10th international conference on World Wide Web*, 406–414. ACM.

Frey, Brendan J., and Delbert Dueck. 2007. Clustering by passing messages between data points. *Science* 315 (5814): 972–976.

Huang, Eric H., Richard Socher, Christopher D. Manning, and Andrew Y. Ng. 2012. Improving word representations via global context and multiple word proto-types. In *Proceedings of the 50th annual meeting of the association for computational linguistics: Long papers*, vol. 1, 873–882. Association for Computational Linguistics.

Iacobacci, Ignacio, Mohammad Taher Pilehvar, and Roberto Navigli. 2015. SensEmbed: Learning sense embeddings for word and relational similarity. In *Proceedings of ACL*, 95–105.

Kågebäck, Mikael, Fredrik Johansson, Richard Johansson, and Devdatt Dubhashi. 2015. Neural context embeddings for automatic discovery of word senses. In *Proceedings of NAACL-HLT*, 25–32.

Kilgarriff, Adam. 2004. How dominant is the commonest sense of a word? In *Text, speech and dialogue: 7th international conference, TSD 2004, Brno, Czech Republic, September 8–11, 2004. Proceedings*, eds. Petr Sojka, Ivan Kopecek, and Karel Pala, 103–111. Berlin, Heidelberg: Springer. ISBN: 978-3-540-30120-2. https://doi.org/10.1007/978-3-540-30120-2_14.

Kleinberg, Jon M. 2003. An impossibility theorem for clustering. In *Advances in neural information processing systems*, 463–470.

Levy, Omer and Yoav Goldberg. 2014. Neural word embedding as implicit matrix factorization. In *Advances in neural information processing systems*, 2177–2185.

Mihalcea, Rada, Timothy Anatolievich Chklovski, and Adam Kilgarriff. 2004. The senseval-3 english lexical sample task. In *Association for computational linguistics*.

Mikolov, Tomas, Kai Chen, Greg Corrado, and Jeffrey Dean. 2013. Efficient estimation of word representations in vector space. arXiv:1301.3781.

Miller, George A. 1995. WordNet: A lexical database for English. *Communications of the ACM* 38 (11): 39–41.

Moro, Andrea and Roberto Navigli. 2015. SemEval-2015 task 13: Multilingual all-words sense disambiguation and entity linking. In *Proceedings of SemEval- 2015*.

Moro, Andrea, Alessandro Raganato, and Roberto Navigli. 2014. Entity linking meets word sense disambiguation: A unified approach. *Transactions of the Association for Computational Linguistics (TACL)* 2: 231–244.

Navigli, Roberto and Simone Paolo Ponzetto. 2010. BabelNet: Building a very large multilingual semantic network. In *Proceedings of the 48th annual meeting of the association for computational linguistics*, 216–225. Association for Computational Linguistics.

Navigli, Roberto, Kenneth C. Litkowski, and Orin Hargraves. 2007. SemEval- 2007 task 07: Coarse-grained english all-words task. In *Proceedings of the 4th international workshop on semantic evaluations. SemEval '07*, 30–35. Prague, Czech Republic: Association for Computational Linguistics.

Neelakantan, Arvind, Jeevan Shankar, Alexandre Passos, and Andrew McCallum. 2015. Efficient non-parametric estimation of multiple embeddings per word in vector space. arXiv:1504.06654.

Pantel, Patrick and Dekang Lin. 2002. Discovering word senses from text. In *Proceedings of the eighth ACM SIGKDD international conference on knowledge discovery and data mining*, 613–619. ACM.

Reisinger, Joseph and Raymond J. Mooney. 2010. Multi-prototype vector-space models of word meaning. In *Human language technologies: The 2010 annual conference of the north american chapter of the association for computational linguistics*, 109–117. Association for Computational Linguistics.

Schütze, Hinrich. 1998. Automatic word sense discrimination. *Computational Linguistics*. 24 (1): 97–123. ISSN: 0891-2017.

Schwenk, Holger. 2004. Efficient training of large neural networks for language modeling. In *2004 IEEE international joint conference on neural networks, 2004. Proceedings*, 4, 3059–3064. IEEE.

Tengi, Randee I. 1998. *Design and implementation of the WordNet lexical database and searching software. WordNet: An electronic lexical database*, ed. Christiane (réd.) Fellbaum, 105. Cambridge: The MIT Press.

Tian, Fei, Hanjun Dai, Jiang Bian, Bin Gao, Rui Zhang, Enhong Chen, and Tie- Yan Liu. 2014. A probabilistic model for learning multi-prototype word embeddings. In *COLING*, 151–160.

White, Lyndon, Roberto Togneri, Wei Liu, and Mohammed Bennamoun. 2018. Finding word sense embeddings of known meaning. In *19th international conference on intelligent text processing and computational linguistics (CICLing)*.

Zipf, George Kingsley. 1945. The meaning-frequency relationship of words. *The Journal of General Psychology* 33 (2): 251–256.

Zipf, G.K. 1949. *Human behavior and the principle of least effort: An introduction to human ecology*. Cambridge: Addison-Wesley Press.

Sentence Representations and Beyond

5

> *A sentence is a group of words expressing a complete thought*
> English Composition and Literature,
> Webster, 1923

Abstract

This chapter discusses representations for larger structures in natural language. The primary focus is on the sentence level. However, many of the techniques also apply to sub-sentence structures (phrases), and super-sentence structures (documents). The three main types of representations discussed here are: unordered models, such as sum of word embeddings; sequential models, such as recurrent neural networks; and structured models, such as recursive autoencoders.

Word Embeddings as a by-product
Many sentence representation methods produce word embeddings as a by-product. These word embeddings are either output embeddings, from the softmax, or input embeddings from a lookup layer.

Initialising input embeddings
It is common (but not ubiquitous) to initialise the input embeddings using pretrained embeddings from one of the methods discussed in Chap. 3, then allow them to be fine-tuned while training the sentence representation method.

It can be argued that the core of true AI, is in capturing and manipulating the representation of an idea. In natural language a sentence (as defined by Webster in the quote above), is such a representation of an idea, but it is not machine

© Springer Nature Singapore Pte Ltd. 2019
L. White et al., *Neural Representations of Natural Language*,
Studies in Computational Intelligence 783,

manipulatable. As such the conversion of sentences to a machine manipulatable representation is an important task in AI research.

All techniques which can represent documents (or paragraphs) by necessity represent sentences as well. A document (or a paragraph), can consist only of a single sentence. Many of these models always work for sub-sentence structures also, like key-phrases. When considering representing larger documents, neural network embedding models directly compete with vector information retrieval models, such as LSI (Dumais et al. 1988), probabilistic LSI (Hofmann and Thomas 2000) and LDA (Blei et al. 2003).

Word Sense Embeddings in Sentence Embeddings
While Chap. 3 was all about sense embeddings, they are unmentioned here. One might think that they would be very useful for sentence embeddings. However, they are not as needful as one might expect. The sense of a word being used is determined by the context. Ideally, it is determined by what the context means. As a sentence embedding is a direct attempt to represent the meaning of such a context, determining the sense of each word within it is not required. Using sense embeddings instead of word embeddings is a valid extension to many of these methods. However it requires performing word sense disambiguation, which as discussed is very difficult.

5.1   Unordered and Weakly Ordered Representations

A model that does not take into account word order cannot perfectly capture the meaning of a sentence. Mitchell and Lapata (2008) give the poignant examples of:

- It was not the sales manager who hit the bottle that day, but the office worker with the serious drinking problem.
- That day the office manager, who was drinking, hit the problem sales worker with a bottle, but it was not serious.

These two sentences have the same words, but in a different structure, resulting in their very different meanings. In practice, however, representations which discard word order can be quite effective.

5.1.1   Sums of Word Embeddings

SOWE is the product of the BOW with an embedding matrix
The reader may recall from Chap. 3, that a word-embedding lookup is the same as a one-hot vector product: $C_{:,w^i} = C \hat{e}_{w^i}$. Similar can be said for sum of word embeddings (SOWE) and bag of words (BOW). For some set of words $\mathcal{W} = \{w_1, \ldots, w_n\}$: the BOW representation is $B_\mathcal{W} = \sum_{w^i \in \mathcal{W}} \hat{e}_{w^i}$; the SOWE representation is $\sum_{w^i \in \mathcal{W}} C_{w^i} = C B_\mathcal{W}$. As with word-embeddings, it is immensely cheaper to calculate this via lookup and sum, rather than via matrix product; except on systems with suitable sparse matrix product tricks.

Classically, in information retrieval, documents have been represented as bags of words (BOW). That is to say a vector with length equal to the size of the vocabulary, with each position representing the count of the number of occurrences of a single word. This is much the same as a *one-hot vector* representing a word, but with every word in the sentence/document counted. The word embedding equivalent is sums of word embeddings (SOWE), and mean of word embeddings (MOWE). These methods, like BOW, lose all order information in the representation. In many cases it is possible to recover a BOW from a much lower dimensional SOWE (White et al. 2016a).

Surprisingly, these unordered methods have been found on many tasks to be extremely well performing, better than several of the more advanced techniques discussed later in this chapter. This has been noted in several works including: White et al. (2015), Ritter et al. (2015) and Wang et al. (2017). It has been suggested that this is because in English there are only a few likely ways to order any given bag of words. It has been noted that given a simple n-gram language model the original sentences can often be recovered from BOW (Horvat and Byrne 2014) and thus also from a SOWE (White et al. 2016b). Thus word-order may not in-fact be as important as one would expect in many natural language tasks, as it is in practice more proscribed than one would expect. That is to say very few sentences with the same word content, will in-practice be able to have it rearranged for a very different meaning. However, this is unsatisfying, and certainly cannot capture fine grained meaning.

The step beyond this is to encode the n-grams into a bag of words like structure. This is a bag of n-grams (BON), e.g. bag of trigrams. Each index in the vector thus represents the occurrence of an n-gram in the text. So It is a good day today, has the trigrams: (It is a),(is a good),(a good day),(good day today). As is obvious for all but the most pathological sentences, recovering the full sentence order from a bag of n-grams is possible even without a language model.

The natural analogy to this with word embeddings might seem to be to find n-gram embeddings by the concatenation of *n* word embeddings; and then to sum these. However, such a sum is less informative than it might seem. As the sum in each concatenated section is equal to the others, minus the edge words.

Instead one should train an n-gram embedding model directly. The method discussed in Chap. 3, can be adapted to use n-grams rather than words as the basic token. This was explored in detail by Li et al. (2017). Their model is based on the skip-gram word embedding method. They take as input an n-gram embedding, and attempt to predict the surrounding n-grams. This reduces to the original skip-gram method for the case of unigrams. Note that the surrounding n-grams will overlap in words (for $n > 1$) with the input n-gram. As the overlap is not complete, this task remains difficult enough to encourage useful information to be represented in the embeddings. Li et al. (2017) also consider training n-gram embeddings as a bi-product of text classification tasks.

5.1.2   Paragraph Vector Models (Defunct)

Window versus Context
It is important to be clear in this section on the difference between the window and the context. The window is the words near the target word. The context (in this context) refers to the larger structure (sentence, paragraph, document) that a representation is attempting to be found for. The window is always a subset of the context. In modelling the context many windows within it will be considered (one per target word). Some works say sentence vector, document vector or paragraph vector. We say context vector as it could be any of the above. In theory it could even be a whole collection of documents.

Le and Mikolov (2014) introduced two models for representing documents of any length by using augmented word-embedding models. The models are called Paragraph Vector Distributed Memory (PV-DM) model, and the Paragraph Vector Distributed Bag of Words model (PV-DBOW). The name Paragraph Vector is a misnomer, it function on texts of any length and has most often (to our knowledge) been applied to documents and sentences rather than any in-between structures. The CBOW and skip-gram models are are extended with an additional context vector that represents the current document (or other super-word structure, such as sentence or paragraph). This, like the word embeddings, is initialised randomly, then trained during the task. Le and Mikolov (2014) considered that the context vector itself must contain useful information about the context. The effect in both cases of adding a context vector is to allow the network to learn a mildly different accusal language model depending on the context. To do this, the context vector would have to learn a representation for the context.

PV Model Implementations
There is a popular third-party implementation of both the paragraph vector models, under the name doc2vec in the python gensim library (Řehůřek et al. 2010), along with many information retrieval vector models such as LDA.

PV-DBOW is an extension of CBOW. The inputs to the model are not only the word-embedding $C :,_{w_j}$ for the words w^j from the window, but also a context-embedding $D_{:,d^k}$ for its current context (sentence, paragraph or document) d^k. The task remains to predict which word was the missing word from the center of the context w^i.

$$P(w^i \mid d^k, w^{i-\frac{n}{2}}, ..., w^{i-1}, w^{i+1}, ..., w^{i+\frac{n}{2}})$$

$$= \text{smax}\left(WD_{:,d^k} + U \sum_{j=i+1}^{j=\frac{n}{2}} \left(C_{:,w^{i-j}} + C_{:,w^{i+j}}\right)\right) \quad (5.1)$$

PV-DM is the equivalent extension for skip-grams. Here the input to the model is not only the central word, but also the context vector. Again, the task remains to predict the other words from the window.

$$P(w^j \mid d^k, w^i) = \left[\text{smax}(WD_{:,d^k} + V C_{:,w^i})\right]_{w_j} \quad (5.2)$$

The results of this work are now considered of limited validity. There were failures to reproduce the reported results in the original evaluations which were on sentiment analysis tasks. These were documented online by several users, including by the second author.[1] A follow up paper, Mesnil et al. (2014) found that reweighed bags of n-grams (Wang et al. 2012) out performed the paragraph vector models. Conversely, Lau et al. (2016) found that on short text-similarity problems, with the right tuning, the paragraph vector models could perform well; however they did not consider the reweighed n-grams of Wang et al. (2012). On a different short text task, White et al. (2015) found the paragraph vector models to significantly be out-performed by SOWE, MOWE, BOW, and BOW with dimensionality reduction. This highlights the importance of rigorous testing against a suitable baseline, on the task in question.

5.2 Sequential Models

The majority of this section draws on the recurrent neural networks (RNN) as discussed in Chap. 2. Every RNN learns a representation of all its input and output in its state. We can use RNN encoders and decoders (as shown in Fig. 5.1) to generate representations of sequences by extracting a coding layer. One can take any RNN encoder, and select one of the hidden state layers after the final recurrent unit (RU) that has processed the last word in the sentence. Similarly for any RNN decoder, one can select any hidden state layer before the first recurrent unit that begins to produce words. For an RNN encoder-decoder, this means selecting the hidden layer from between. This was originally considered in Cho et al. (2014b), when using a machine translation RNN, to create embeddings for the translated phrases. Several other RNNs have been used in this way since.

---

[1] https://groups.google.com/forum/#!msg/word2vec-toolkit/Q49FIrNOQRo/DoRuBoVNFb0J.

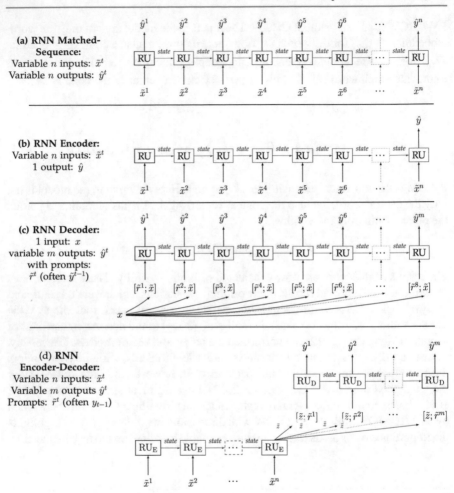

Fig. 5.1 The unrolled structure of an RNN for use in **a** Matched-sequence **b** Encoding, **c** Decoding and **d** Encoding-Decoding (sequence-to-sequence) problems. RU is the recurrent unit – the neural network which reoccurs at each time step. (Repeated from Fig. 2.1)

5.2.1 VAE and Encoder-Decoder

Bowman et al. (2016b) presents an extension on this notion, where in-between the encode and the decode stages there is a variational autoencoder (VAE). This is shown in Fig. 5.2. The variational autoencoder (Kingma and Welling 2014) has been demonstrated to have very good properties in a number of machine learning applications: they are able to work to find continuous latent variable distributions over arbitrary likelihood functions (such as in the neural network); and are very fast to train. Using the VAE, it is hoped that a better representation can be found for the sequence of words in the input and output.

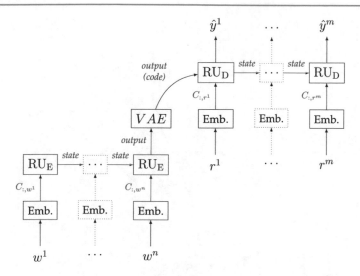

Fig. 5.2 The VAE plus encoder-decoder of Bowman et al. (2016b). Note that during training, $\hat{y}^i = w^i$, as it is an autoencoder model. As is normal for encoder-decoders the prompts are the previous output (target during training, predicted during testing): $r^i = y^{i-1}$, with $r^1 = y_0 = \text{<EOS>}$ being a pseudo-token marker for the end of the string. The Emb. step represents the embedding table lookup. In the diagrams for Chap. 3 we showed this as as a table but just as a block here for conciseness

Bowman et al. (2016b) trained the network as encoder-decoder reproducing its exact input. They found that short syntactically similar sentences were located near to each other according to this space, further to that, because it has a decoder, it can generate these nearby sentences, which is not possible for most sentence embedding methods.

Interestingly, they use the VAE output, i.e. the *code*, only as the state input to the decoder. This is in-contrast to the encoder-decoders of Cho et al. (2014b), where the *code* was concatenated to the input at every timestep of the decoder. Bowman et al. (2016b) investigated such a configuration, and found that it did not yield an improvement in performance.

5.2.2 Skip-Thought

Kiros et al. (2015) draws inspiration from the works on acausal language modelling, to attempt to predict the previous and next sentence. Like in the acausal language modelling methods, this task is not the true goal. Their true goal is to capture a good representation of the current sentence. As shown in Fig. 5.3 they use an encoder-decoder RNN, with two decoder parts. One decoder is to produce the previous

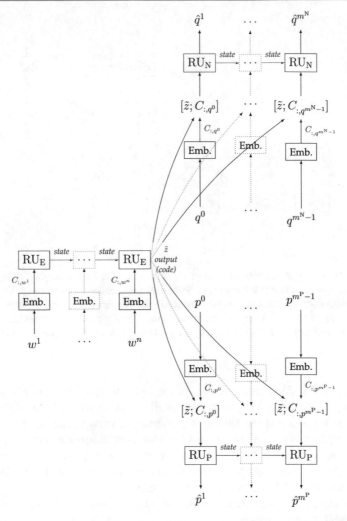

Fig. 5.3 The skip-thought model (Kiros et al. 2015). Note that for the next and previous sentences respectively the outputs are $\hat{q}^i$ and $\hat{p}^i$, and the prompts are q^{i-1} and p^{i-1}. As there is no intent to use the decoders after training, there is no need to worry about providing an evaluation-time prompt, so the prompt is always the previous word. $p^0 = p^{m^P} = q^0 = q^{m^q} = $ <EOS> being a pseudo-token marker for the end of the string. The input words are w^i, which come from the current sentence. the Emb. steps represents the look-up of the embedding for the word

sentence. The encoder part takes as it's input is the current sentence, and produces as its output the code, which is input to the decoders. The other decoder is to produce the next sentence. As described in Sect. 2.2.3, the prompt used for the decoders includes the previous word, concatenated to the code (from the encoder output).

That output code is the representation of the sentence.

5.3   Structured Models

Parsers
There are many high-quality parsing libraries available. The most well known is the Stanford CoreNLP library (Manning et al. 2014) for java. It has an interactive web-demo at http://corenlp.run/, which was used to produce Figs. 5.4 and 5.5. NLTK (Bird et al. 2009) contains several different parsers, including a binding to CoreNLP parsers. The newer spaCy library (Honnibal et al. 2015) for python, presently only features a dependency parser.

The sequential models are limited to processed the information as a series of timesteps one after the other. They processes sentences as ordered lists of words. However, the actual structure of a natural language is not so simple. Linguists tend to break sentences down into a tree structure. This is referred to as parsing. The two most common forms are constituency parse trees, and dependency parse trees. Examples of each are shown in Figs. 5.4 and 5.5. It is beyond the scope of this book to explain the precise meaning of these trees, and how to find them. The essence is that these trees represent the structure of the sentence, according to how linguists believe sentences are processed by humans.

The constituency parse breaks the sentence down into parts such as noun phrase (NP) and verb phrase (VP), which are in turn broken down into phrases, or (POS tagged) words. The constituency parse is well thought-of as a hierarchical breakdown of a sentence into its parts. Conversely, a dependency parse is better thought of as a set of binary relations between head-terms and their dependent terms. These structures are well linguistically motivated, so it makes sense to use them in the processing of natural language.

We refer here to models incorporating tree (or graph) structures as structural models. Particular variations have their own names, such as recursive neural networks (RvNN), and recursive autoencoders (RAE). We use the term structural model as an all encompassing term, and minimise the use of the easily misread terms: recursive versus recurrent neural networks. A sequential model (an RNN) is a particular case of a structural model, just as a linked list is a particular type of tree. However, we will exclude sequential models them this discussion except where marked.

The initial work on structural models was done in the thesis of Socher (2014). It builds on the work of Goller and Kuchler (1996) and Pollack (1990), which present back-propagation through structure. Back-propagation can be applied to networks of any structure, as the chain-rule can be applied to any differentiable equation to find its derivative. Structured networks, like all other networks, are formed by the composition of differentiable functions, so are differentiable. In a normal network

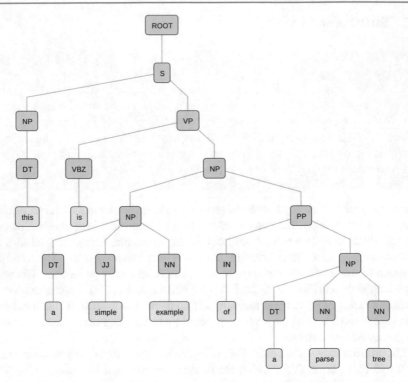

Fig. 5.4 A constituency parse tree for the sentence: `This is a simple example of a parse tree`. In this diagram the leaf nodes are the input words, their intimidate parents are their POS tags, and the other nodes with multiple children represent sub-phrases of the sentence, for example NP is a Noun Phrase

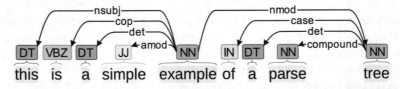

Fig. 5.5 A dependency parse tree for the sentence `This is a simple example of a parse tree`, This flattened view may be misleading. `example` is at the peak of the tree, with direct children being: `this,is,a,simple`, and `tree`. `tree` has direct children being: `of,a`, and `parse`

the same composition of functions is used for all input cases, whereas in a structured network it is allowed to vary based on the inputs. This means that structuring a network according to its parse tree is possible.

Machine learning frameworks for structural models

Structural networks cannot be easily defined in most static neural network libraries, such as TensorFlow. These implementations function by defining a single computational graph that is used to process each training/test case. The same graph is used for each input. By definition, the structure of the network differs from training/test case to training/test case. Technically the same problems apply to RNNs, as each case can have a different number of inputs. This is normally worked around by defining the network graph for the longest input to be considered, then padding all the inputs to this length, and ensuring that the padding does not interfere with the gradient updates. The equivalent tricks for structured networks are significantly more complex. The exception to this is of-course via dynamic components to the static frameworks (which TensorFlow and other such frameworks certainly do have). Even in a dynamic framework it remains a non-trivial task to implement these networks.

Implementing Back-propagation through structure

Conceptually, back-propagation through structure is not significantly more complex than back-propagation through time. However, in practice it is a very difficult algorithm to get right. It is very important to test for correctness using gradient checks, as it is easy to make a mistake and end-up with models that seem to work ok, but are actually crippled due to some mistake in the coding. Unfolding recursive autoencoders are particularly difficult, as the gradient must be propagated from all leaves. And output interior nodes cannot have their gradients calculated until the gradients of their children are calculated. The solution to this is to process the node gradient calculation using a priority queue, where the priority is set by the depth of the node. Thus ensuring that all children are processed before their parents.

5.3.1   Constituency Parse Tree (Binary)

Tree structured networks work by applying a recursive unit (which we will call RV) function across pairs (or other groups) of the representations of the lower levels, to produce a combined representation. The network structure for an input of binary tree structured text is itself a binary tree of RVs. Each RV (i.e. node in the graph) can be defined by the composition function:

$$f^{\mathrm{RV}}(\tilde{u}, \tilde{v}) = \varphi\left(\begin{bmatrix} S & R \end{bmatrix} \begin{bmatrix} \tilde{u} \\ \tilde{v} \end{bmatrix} + \tilde{b} \right) \tag{5.3}$$

$$= \varphi\left(S\tilde{u} + R\tilde{v} + \tilde{b} \right) \tag{5.4}$$

where $\tilde{u}$ and $\tilde{v}$ are the left and right substructures embeddings (word embeddings at the leaf node level), and S and R are the matrices defining how the left and right children's representations are to be combined.

This is a useful form as all constituency parse trees can be converted into binary parse trees, via left-factoring or right factoring (adding new nodes to the left or right to take some of the children). This is sometimes called binarization, or putting them into Chomsky normal form. This form of structured network has been used in many words, including (Socher et al. 2010, 2011a,b,c; Zhang et al. 2014). Notice that S and R matrices are shared for all RVs, so all substructures are composed in the same way, based only on whether they are on the left, or the right.

5.3.2 Dependency Tree

The dependency tree is the other commonly considered parse-tree. Structured networks based upon the dependency tree have been used by Socher et al. (2014), Iyyer et al. (2014a, b). In these works rather than a using composition matrix for left-child and right-child, the composition matrix varies depending on the type of relationship of between the head word and its child. Each dependency relationship type has its own composition matrix. That is to say there are distinct composition matrices for each of nsub, det, nmod, case etc. This allows for multiple inputs to a single head node to be distinguished by their relationship, rather than their order. This is important for networks using a dependency parse tree structure as the relationship is significant, and the structure allows a node to have any number of inputs.

Extended example
The full example for the $f^{\mathrm{RV}}(9)$ from Eq. 5.5 is:

$$
\begin{aligned}
f^{\mathrm{RV}}(9) = \varphi(W^{\mathrm{head}} C_{:,\mathrm{tree}} \\
+ W^{\mathrm{compound}}(W^{\mathrm{head}} C_{:,\mathrm{parse}} + \tilde{b}) \\
+ W^{\mathrm{det}}(W^{\mathrm{head}} C_{:,\mathrm{a}} + \tilde{b}) \\
+ W^{\mathrm{case}}(W^{\mathrm{head}} C_{:,\mathrm{of}} + \tilde{b}) \\
+ \tilde{b})
\end{aligned}
$$

This in turn would be composed as part of $f^{\mathrm{RV}}(5)$ for the whole tree headed by $w^5 = \mathtt{example}$. The output of each RV is a representation of that substructure.

No gates No long-term memory
We note that a limitation of most structural models, compared to the sequential RNNs, is their lack of explicit gating on memory (e.g. as in GRU and LSTM). Any given path down a tree can be looked at as a simple RNN comprised only of basic recurrent units. However, these paths are much shorter (being the logarithm of) than the full sequential length of the sentence, which offsets the need for such gating. Recall that the gating is to provide the longer short term memory.

Consider a function $\pi(i, j)$ which returns the relationship between the head word at position i and the child word at position j. For example, using the tree shown in Fig. 5.5, which has $w^8 = \texttt{parse}$ and $w^9 = \texttt{tree}$ then $\pi(8, 9) = \texttt{compound}$. This is used to define the composed representation for each RV:

$$f^{RV}(i) = \varphi \left(W^{\text{head}} C_{:,w^i} + \sum_{j \in \text{children}(i)} W_{\pi(i,j)} \, f_{RV}(j) + \tilde{b} \right) \qquad (5.5)$$

Here $C_{:,w^i}$ is the word embedding for w^i, and W_{head} encodes the contribution of the headword to the composed representation. Similarly, $W_{\pi(i,j)}$ encodes the contribution of the child words. Note that the terminal case is just $f_{RV}(i) = \varphi \left(W^{\text{head}} C_{:,w^i} + \tilde{b} \right)$ when a node i has no children. This use of the relationship to determine the composition matrix, increases both the networks expressiveness, and also handles the non-binary nature of dependency trees.

A similar technique could be applied to constituency parse trees. This would be using the part of speech (e.g. VBZ, NN) and phrase tags (e.g. NP, VP) for the sub-structures to choose the weight matrix. This would, however, lose the word-order information when multiple inputs have the same tag. This would be the case, for example, in the right-most branch shown in Fig. 5.4, where both `parse` and `tree` have the NN POS tag, and thus using only the tags, rather than the order would leave `parse tree` indistinguishable from `tree parse`. This is not a problem for the dependency parse, as word relationships unambiguously correspond to the role in the phrase's meaning. As such, allowing the dependency relationship to define the mathematical relationship, as encoded in the composition matrix, only enhances expressibility.

For even greater capacity for the inputs to control the composition, would be to allow every word be composed in a different way. This can be done by giving the child nodes there own composition matrices, to go with there embedding vectors. The composition matrices encode the relationship, and the operation done in the composition. So not only is the representation of the (sub)phrase determined by a relationship between its constituents (as represented by their embeddings), but the nature of that relationship (as represented by the matrix) is also determined by those same constituents. In this approach at the leaf-nodes, every word not only has a word vector, but also a word matrix. This is discussed in Sect. 5.4.

5.3.3 Parsing

The initial work for both contingency tree structured networks (Socher et al. 2010) and for dependency tree structured networks (Stenetorp 2013) was on the creation of parsers. This is actually rather different to the works that followed. In other works the structure is provided as part of the input (and is found during preprocessing). Whereas a parser must induce the structure of the network, from the unstructured input text. This is simpler for contingency parsing, than for dependency parsing.

The finer detail of parsing
Parsing is one of the most well studied problems in computational linguistics. Presented here is only the highest level overview. For more details on this, we recommend consulting the source materials. Ideally, with reference to a good traditional (that is to say non-neural network based) NLP textbook, such as: Manning et al. (1999).

Getting the Embeddings out of the Parser
The implementation of Socher et al. (2010), is publicly available. However, it does not export embeddings. It is nested deep inside the Stanford Parser, and thus accessing the embeddings is not at all trivial.

When creating a binary contingency parse tree, any pair of nodes can only be merged if they are adjacent. The process described by Socher et al. (2010), is to consider which nodes are to be composed into a higher level structure each in turn. For each pair of adjacent nodes, an RV can be applied to get a merged representation. A linear scoring function is also learned, that takes a merged representation and determines how good it was. This is trained such that correct merges score highly. Hinge loss is employed for this purpose. The Hinge loss function works on similar principles to negative sampling (see the motivation given in Sect. 3.4.2). Hinge loss is used to cause the merges that occur in the training set to score higher than those that do not. To perform the parse, nodes are merged; replacing them with their composed representation; and the new adjacent pairing score is then recomputed. Socher et al. (2010) considered both greedy, and dynamic programming search to determine the order of composition, as well as a number of variants to add additional information to the process. The dependency tree parser extends beyond this method.

Dependency trees can have child-nodes that do not correspond to adjacent words (non-projective language). This means that the parser must consider that any (unlinked) node be linked to any other node. Traditional transition-based dependency parsers function by iteratively predicting links (transitions) to add to the structure based on its current state. Stenetorp (2013) observed that a composed representation similar to Eq. (5.3), was an ideal input to a softmax classifier that would predict the next link to make. Conversely, the representation that is suitable for predicting the next link to make, is itself a composed representation. Note, that Stenetorp (2013) uses the same matrices for all relationships (unlike the works discussed in Sect. 5.3.2). This is required, as the relationships must be determined from the links made, and thus are not available before the parse. Bowman et al. (2016a), presents a work an an extension of the same principles, which combines the parsing step with the processing of the data to accomplish some task, in their case detecting entailment.

5.3.4   Recursive Autoencoders

Application to image retrieval
An interesting application of structured networks was shown in Socher et al. (2014). A dependency tree structured network was trained on a language modelling task (not as a recursive autoencoder, although that would also have been a valid option). Then, separately a convolutional neural network was trained to produce a vector output of the same dimensionality – an image embedding – such that its distance to its caption's composed vector representation was minimised. The effect of this was that images and their captions are projected to a common vector space. This allowed for smart image retrieval, from descriptions, by having a set of all images, and storing their embedding representations. Then for any query, the sentence embedding can be found and the vector space of images can be searched for the nearest. The reverse is not generally as useful, as one can't reasonably store all possible captions describing an image, so as to be able to search for the best one for a user provided image. This process of linking a sequence representation to an image embedding is not restricted to structured networks, and can be done with any of the representation methods discussed in this chapter. Further, as discussed in Sect. 3.6 it can also be done using pretrained embedding on both sides through (kernel) CCA.

Unfolding RAE implementation
The implementation, and a pretrained model, of the unfolding recursive autoencoder of Socher et al. (2011a) is available online at https://tinyurl.com/URAE2011. It is easy to use as a command-line Matlab script to generate embeddings.

Recursive autoencoders are autoencoders, just as the autoencoder discussed in Sect. 1.5.2, they reproduce their input. It should be noted that unlike the encoder-decoder RNN discussed in Sect. 5.2.1, they cannot be trivially used to generate natural language from an arbitrary embeddings, as they require the knowledge of the tree structure to unfold into. Solving this would be the inverse problem of parsing (discussed in Sect. 5.3.3).

The models presented in Socher et al. (2011a) and Iyyer et al. (2014a) are unfolding recursive autoencoders. In these models an identical inverse tree is defined above the highest node. The loss function is the sum of the errors at the leaves, i.e. the distance in vector space between the reconstructed words embeddings and the input word-embeddings. This was based on a simpler earlier model: the normal (that is to say, not unfolding) recursive autoencoder.

The normal recursive autoencoder, as used in Socher et al. (2011c) and Zhang et al. (2014) only performs the unfolding for a single node at a time during training. That means that it assesses how well each merge can individually be reconstructed, not the success of the overall reconstruction. This per merge reconstruction has a loss function based on the difference between the reconstructed embeddings and the inputs embeddings. Note that those inputs/reconstructions are not word embeddings: they are the learned merged representations, except when the inputs happen to be leaf node. This single unfold loss covers the auto-encoding nature of each merge; but does not give any direct assurances of the auto-encoding nature of the whole structure. However, it should be noted that while it is not trained for, the reconstruction components (that during training are applied only at nodes) can nevertheless be applied recursively from the top layer, to allow for full reconstruction.

5.3.4.1 Semi-supervised

In the case of all these autoencoders, except (Iyyer et al. 2014a), a second source of information is also used to calculate the loss during training. The networks are being simultaneously trained to perform a task, and to regenerate their input. This is often considered as semi-supervised learning, as unlabelled data can be used to train the auto-encoding part (unsupervised) gaining a good representation, and the labelled data can be used to train the task output part (supervised) making that representation useful for the task. This is done by imposing an additional loss function onto the output of the central/top node.

- In Socher et al. (2011c) this was for sentiment analysis.
- In Socher et al. (2011a) this was for paraphrase detection.
- In Zhang et al. (2014) this was the distance between embeddings of equivalent translated phrases of two RAEs for different languages.

The reconstruction loss and the supervised loss can be summed, optimised in alternating sequences, or the reconstructed loss can be optimised first, then the labelled data used for fine-tuning.

Sequential models are often preferred to structural models
Sequential (RNN) models are much more heavily researched than structural models. They have better software libraries, are easier to implement, and have more known "tricks" (like gates and attention). In theory it is possible for a sequential model (with sufficiently deep and wide RUs) to internalise the connections that a structural model would possess. While structural models are theoretically nicer from a linguistics standpoint, pragmatically they are the last resort in modelling. When attempting to find a useful representation of a sentence for a task, one should first try a sum of word embeddings with a simple network on-top, then a sequential model (based on LSTM or GRU), and only if these fail then try a structured model. Arguably before using any neural network approach at all, one should eliminate bag of words, bag of n-grams, the dimensionality reduced version of those bags, and also eliminate LSI and LDA as inputs for the task.

5.4 Matrix Vector Models

5.4.1 Structured Matrix Vector Model

Socher et al. (2012) proposed that each node in the graph should define not only a vector embedding, but a matrix defining how it was to be combined with other nodes. That is to say, each word and each phrase has both an embedding, and a composition matrix.

Consider this for binary constituency parse trees. The composition function is as follows:

$$f^{\text{RV}}(\tilde{u}, \tilde{v}, U, V) = \varphi\left([S\ R][U\tilde{v}; V\tilde{u}] + \tilde{b}\right) \tag{5.6}$$

$$= \varphi\left(S\, U\tilde{v} + R\, V\tilde{u} + \tilde{b}\right) \tag{5.7}$$

$$F^{\text{RV}}(U, V) = W\,[U; V] = W_l U + W_r V \tag{5.8}$$

f^{RV} gives the composed embedding, and F^{RV} gives the composing matrix. The S and R represent the left and right composition matrix components that are the same for all nodes (regardless of content). The U and V represent the per word/node child composition matrix components. We note that S and R could, in theory, be rolled in to U and R as part of the learning. The $\tilde{u}$ and $\tilde{v}$ represent the per word/node children embeddings, and W represents the method for merging two composition matrices.

We note that one can define increasingly complex and powerful structured networks along these lines; though one does run the risk of very long training times and of over-fitting.

5.4.2 Sequential Matrix Vector Model

A similar approach, of capturing a per word matrix, was used on a sequential model by Wang et al. (2017). While sequential models are a special case of structured models, it should be noted that unlike the structured models discussed prior, this matrix vector RNN features a gated memory. This matrix-vector RNN is an extension of the GRU discussed in Chap. 2, but without a reset gate.

In this sequential model, advancing a time step, is to perform a composition. This composition is for between the input word and the (previous) state. Rather than directly between two nodes in the network as in the structural case. It should be understood that composing with the state is not the same as composing the current input with the previous input. But rather as composing the current input with all previous inputs (though not equally).

As depicted in Fig. 5.6 each word, w^t is represented by a word embedding $\tilde{x}^t$ and matrix: $\tilde{X}^{w^t}$, these are the inputs at each time step. The network outputs and states are the composed embedding $\hat{y}^t$ and matrix Y^t.

> **Remember:**
> The product of a matrix with a concatenated vector, is the same as the sum of the two blocks of that matrix each multiplied by the blocks of that vector.

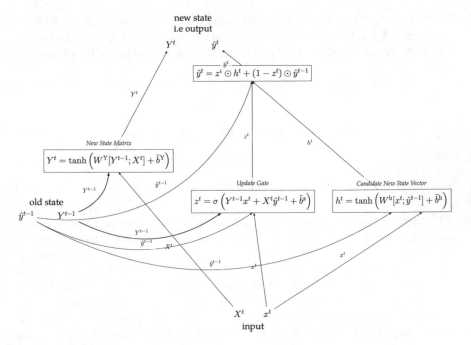

Fig. 5.6 A Matrix Vector recurrent unit

$$h^t = \tanh\left(W^h[x^t; \hat{y}^{t-1}] + \tilde{b}^h\right) \tag{5.9}$$

$$z^t = \sigma\left(Y^{t-1}x^t + X^t\hat{y}^{t-1} + \tilde{b}^z\right) \tag{5.10}$$

$$\hat{y}^t = z^t \odot h^t + (1 - z^t) \odot \hat{y}^{t-1} \tag{5.11}$$

$$Y^t = \tanh\left(W^Y[Y^{t-1}; X^t] + \tilde{b}^Y\right) \tag{5.12}$$

The matrices W^h, W^Y and the biases $\tilde{b}^h$, $\tilde{b}^z$, $\tilde{b}^Y$ are shared across all time steps/compositions. W^Y controls how the next state-composition Y^t matrix is generated from its previous value and the input composition matrix, X^t; W^h similarly controls the value of the candidate state-embedding h^t.

h^t is the candidate composed embedding (to be output/used as state). z^t is the update gate, it controls how much of the actual composed embedding ($\hat{y}^t$) comes from the candidate h^t and how much comes from the previous value ($\hat{y}^{t-1}$). The composition matrix Y^t (which is also part of the state/output) is not gated.

Notice, that the state composition matrix Y^{t-1} is only used to control the gate z^t, not to directly affect the candidate composited embedding h^t. Indeed, in fact one can note that all similarity to the structural method of Socher et al. (2012) is applied in the gate z^t. The method for calculating h^t is similar to that of a normal RU.

The work of Wang et al. (2017), was targeting short phrases. This likely explains the reason for not needing a forget gates. The extension is obvious, and may be beneficial when applying this method to sentences.

5.5 Conclusion, on Compositionality

It is tempting to think of the structured models as compositional, and the sequential models as non-compositional. However, this is incorrect.

The compositional nature of the structured models is obvious: the vector for a phrase is composed from the vectors of the words that the phrase is formed from.

Sequential models are able to learn the structures. For example, learning that a word from n time steps ago is to be remembered in the RNN state, to then be optimally combined with the current word, in the determination of the next state. This indirectly allows the same compositionality as the structured models. It has been shown that sequential models are indeed in-practice able to learn such relationships between words (White et al. 2017). More generally as almost all aspects of language have some degree of compositionality, and sequential models work very well on most language tasks, this implicitly shows that they have sufficient representational capacity to learn sufficient degrees of compositional processing to accomplish these tasks.

In fact, it has been suggested that even some unordered models such as sum of word embeddings are able to capture some of what would be thought of as compositional information. Ritter et al. (2015) devised a small corpus of short sentences describing containing relationships between the locations of objects. The task and dataset was constructed such that a model must understand some compositionality, to be able to classify which relationships were described. Ritter et al. (2015) tested several sentence representations as the input to a naïve Bayes classifier being trained to predict the relationship. They found that when using sums of high-quality word embeddings as the input, the accuracy not only exceeded the baseline, but even exceeded that from using representation from a structural model. This suggests that a surprising amount of compositional information is being captured into the embeddings; which allows simple addition to be used as a composition rule. Though being ignorant of word order does mean it certainly couldn't be doing so perfectly, however the presence of other words my be surprisingly effective hinting at the word order (White et al. 2016b), thus allow for more apparently compositional knowledge to be encoded than is expected.

To conclude, the compositionality capacity of many models is not as clear cut as it may initially seem. Further to that the requirement for a particular task to actually handle compositional reasoning is also not always present, or at least not always a significant factor in practical applications. We have discussed many models in this section, and their complexity varies significantly. They range from the very simple sum of word embeddings all the way to the the structured matrix models, which are some of the more complicated neural networks ever proposed.

References

Bird, Steven, Ewan Klein, and Edward Loper. 2009. Natural language processing with Python. O'Reilly Media, Inc.

Blei, David M., Andrew Y. Ng, and Michael I. Jordan. 2003. Latent Dirichlet allocation. *The Journal of Machine Learning Research* 3: 993–1022.

Bowman, Samuel R., Jon Gauthier, Abhinav Rastogi, Raghav Gupta, Christopher D. Manning, and Christopher Potts. 2016a. A fast unified model for parsing and sentence understanding. arXiv:1603.06021.

Bowman, Samuel R., Luke Vilnis, Oriol Vinyals, Andrew M. Dai, Rafal Jozefowicz, and Samy Bengio. 2016b. Generating sentences from a continuous space. In *International conference on learning representations (ICLR) Workshop*.

Cho, Kyunghyun, Bart van Merrienboer, Caglar Gulcehre, Dzmitry Bahdanau, Fethi Bougares, Holger Schwenk, and Yoshua Bengio. 2014. Learning phrase representations using rnn encoder-decoder for statistical machine translation. In *Proceedings of the 2014 conference on empirical methods in natural language processing (EMNLP)*, 1724–1734. Doha, Qatar: Association for Computational Linguistics.

Dumais, Susan T., George W. Furnas, Thomas K. Landauer, Scott Deerwester, and Richard Harshman. 1988. Using latent semantic analysis to improve access to textual information. In *Proceedings of the SIGCHI conference on Human factors in computing systems*, 281–285. ACM.

Goller, Christoph and Andreas Kuchler. 1996. Learning task-dependent distributed representations by back propagation through structure. In *IEEE international conference on neural networks, 1996*, vol. 1, 347–352. IEEE.

Hofmann, Thomas. 2000. Learning the similarity of documents: An information geometric approach to document retrieval and categorization. In *Advances in neural information processing systems*, 914–920.

Honnibal, Matthew and Mark Johnson. 2015. An improved non-monotonic transition system for dependency parsing. In *Proceedings of the 2015 conference on empirical methods in natural language processing*, 1373–1378. Lisbon, Portugal: Association for Computational Linguistics.

Horvat, Matic and William Byrne. 2014. A graph-based approach to string regeneration. In *EACL*, 85–95.

Iyyer, Mohit, Jordan Boyd-Graber, and Hal Daumé III. 2014a. Generating sentences from semantic vector space representations. In *NIPS workshop on learning semantics*.

Iyyer, Mohit, Jordan Boyd-Graber, Leonardo Claudino, Richard Socher, and Hal Daumé III. 2014b. A neural network for factoid question answering over para-graphs. In *Proceedings of the 2014 conference on empirical methods in natural language processing (EMNLP)*, 633–644.

Kingma, D. P. and M. Welling. 2014. Auto-encoding variational bayes. In *The international conference on learning representations (ICLR)*. arXiv:1312.6114 [stat.ML].

Kiros, Ryan, Yukun Zhu, Ruslan Salakhutdinov, Richard S. Zemel, Antonio Torralba, Raquel Urtasun, and Sanja Fidler. 2015. Skip-thought vectors. In *CoRR*. arXiv:1506.06726.

Lau, Jey Han, and Timothy Baldwin. 2016. An empirical evaluation of doc2vec with practical insights into document embedding generation. In *ACL*, 78.

Le, Quoc and Tomas Mikolov. 2014. Distributed representations of sentences and documents. In *Proceedings of the 31st international conference on machine learning (ICML-14)*, 1188–1196.

Li, Bofang, Tao Liu, Zhe Zhao, Puwei Wang, and Xiaoyong Du. 2017. Neural bag-of-ngrams. In *AAAI*, 3067–3074.

Manning, C.D. and H. Schütze. 1999. Foundations of statistical natural language processing. Cambridge: MIT Press. ISBN: 9780262133609.

Manning, Christopher D., Mihai Surdeanu, John Bauer, Jenny Finkel, Steven J. Bethard, and David McClosky. 2014. The Stanford CoreNLP natural language processing toolkit. In *Association for computational linguistics (ACL) system demonstrations*, 55–60.

Mesnil, Grégoire, Tomas Mikolov, Marc'Aurelio Ranzato, and Yoshua Bengio. 2014. Ensemble of generative and discriminative techniques for sentiment analysis of movie reviews. arXiv:1412.5335.

Mitchell, Jeff and Mirella Lapata. 2008. Vector-based models of semantic composition. In *ACL*, 236–244.

Pollack, Jordan B. 1990. Recursive distributed representations. *Artificial Intelligence*, 46 (1): 77–105. ISSN: 0004-3702. https://doi.org/10.1016/0004-3702(90)90005-K.

Řehůřek, Radim and Petr Sojka. 2010. Software framework for topic modelling with large corpora. English. In *Proceedings of the LREC 2010 workshop on new challenges for NLP frameworks*, 45–50. Valletta, Malta: ELRA. http://is.muni.cz/publication/884893/en

Ritter, Samuel, Cotie Long, Denis Paperno, Marco Baroni, Matthew Botvinick, and Adele Goldberg. 2015. Leveraging preposition ambiguity to assess compositional distributional models of semantics. In *The fourth joint conference on lexical and computational semantics*.

Socher, Richard. 2014. Recursive deep learning for natural language processing and computer vision. Ph.D. thesis. Stanford University.

Socher, Richard, Christopher D. Manning, and Andrew Y. Ng. 2010. Learning continuous phrase representations and syntactic parsing with recursive neural networks. In *Proceedings of the NIPS-2010 deep learning and unsupervised feature learning workshop*, 1–9.

Socher, Richard, Eric H. Huang, Jeffrey Pennington, Andrew Y. Ng, and Christopher D. Manning. 2011a. Dynamic pooling and unfolding recursive autoencoders for paraphrase detection. In *Advances in neural information processing systems*, 24.

Socher, Richard, Cliff C Lin, Chris Manning, and Andrew Y Ng. 2011b. Parsing natural scenes and natural language with recursive neural networks. In *Proceedings of the 28th international conference on machine learning (ICML-11)*, 129–136.

Socher, Richard, Jeffrey Pennington, Eric H. Huang, Andrew Y. Ng, and Christopher D. Manning. 2011c. Semi-supervised recursive autoencoders for predicting sentiment distributions. In *Proceedings of the 2011 conference on empirical methods in natural language processing (EMNLP)*.

Socher, Richard, Brody Huval, Christopher D. Manning, and Andrew Y. Ng. 2012. Semantic compositionality through recursive matrix-vector spaces. In *Proceedings of the 2012 joint conference on empirical methods in natural language processing and computational natural language learning*, 1201–1211. Association for Computational Linguistics.

Socher, Richard, Andrej Karpathy, Quoc V. Le, Christopher D. Manning, and Y.Ng Andrew. 2014. Grounded compositional semantics for finding and describing images with sentences. *Transactions of the Association for Computational Linguistics* 2: 207–218.

Stenetorp, Pontus. 2013. Transition-based dependency parsing using recursive neural networks. In *Deep learning workshop at the, 2013. Conference on neural information processing systems (NIPS)*. Nevada, USA: Lake Tahoe.

Wang, Sida and Christopher D. Manning. 2012. Baselines and bigrams: Simple, good sentiment and topic classification. In *Proceedings of the 50th annual meeting of the association for computational linguistics: Short papers*, vol. 2, 90–94. Association for Computational Linguistics.

Wang, Rui, Wei Liu, and Chris McDonald. 2017. A matrix-vector recurrent unit model for capturing compositional semantics in phrase embeddings. In *International conference on information and knowledge management*.

White, Lyndon, Roberto Togneri, Wei Liu, and Mohammed Bennamoun. 2015. How well sentence embeddings capture meaning. In *Proceedings of the 20th Australasian document computing symposium. ADCS '15*, 9:1–9:8. Parramatta, NSW, Australia: ACM. ISBN: 978-1-4503-4040-3, https://doi.org/10.1145/2838931.2838932.

White, Lyndon, Roberto Togneri, Wei Liu, and Mohammed Bennamoun. 2016a. Generating bags of words from the sums of their word embeddings. In *17th international conference on intelligent text processing and computational linguistics (CICLing)*.

White, Lyndon, Roberto Togneri, Wei Liu, and Mohammed Bennamoun. 2016b. Modelling sentence generation from sum of word embedding vectors as a mixed integer programming problem. In *IEEE international conference on data mining: High dimensional data mining workshop (ICDM: HDM)*. https://doi.org/10.1109/ICDMW.2016.0113.

White, L., R. Togneri, W. Liu, and M. Bennamoun. 2017. Learning distributions of meant color. arXiv:1709.09360 [cs.CL].

Zhang, Jiajun, Shujie Liu, Mu Li, Ming Zhou, and Chengqing Zong. 2014. Bilingually constrained phrase embeddings for machine translation. In *ACL*.

Conclusion

6

The key to artificial intelligence has always been the
representation. You and I are streaming data engines

Jeff Hawkins 2012

Abstract

In this book we have introduced methods for finding representations of natural language. A special focus of this book is on neural networks and related technologies. Neural networks elegantly produce useful representations as by-products, when applied to NLP tasks. We have introduced recent advances in neural networks, in particular, various forms of recurrent neural networks, and have covered techniques for working with words, word senses, and larger structures such as phrases, sentences and documents.

Some final reminders

This section has some final reminders, of points made earlier, and some final remarks not made elsewhere.

Reminder: Gradient Checks

Back-propagation is a very easy algorithm to mess up, and it is hard to tell when you do mess it up, because it will often still work reasonably well. Gradient checks, either with finite-differencing, or with more sophisticated techniques (e.g. dual-numbers), is a quick and easy way to check the correctness of your implementations.

Chapter 1 introduced the idea of machine learning, which allows us to train a system using examples of the desired outputs for given inputs. The system can learn to

© Springer Nature Singapore Pte Ltd. 2019
L. White et al., *Neural Representations of Natural Language*,
Studies in Computational Intelligence 783,

determine the outputs for inputs that were never seen during training. A key feature of modern machine learning is the decreased reliance on hand-engineered features to represent their inputs. The learnt embeddings and vector representations which are the main topic of this book, are just such automatically derived features.

Hyperparameters matter
Neural networks have many hyper parameters, from hidden layer-size, to the regularisation penalty, to the number of training iterations. Choosing the right ones can significantly impact performance.

Chapter 2 showed how recurrent neural networks allowed working with inputs of varying lengths – such as natural language senses. It discussed the various types of recurrent networks as characterised by their recurrent units, including both GRU and LSTM. As well as the common RNN structures, including: the encoder for producing a fixed-size output (such as a classification) from a varying sized input; the decoder for generating a varying sized output from a fixed size input; and the encoder-decoder for when the input and output both vary in size. These models are very practical for natural language processing applications.

Preprocessing Matters
There is a large number of preprocessing tricks that can be employed when processing text. Options include removing stop-words, rare words, and punctuation, as well as removing or replacing numbers and dates. Different options for tokenization exist with regard to splitting up contractions and other factors. Often, it is good to convert all the text to lower-case. One can even lemmatize or stem every word occurrence. What is useful depends on the task.

Chapter 3 discussed how we can find and use representations of words. This is one of the most important ideas in machine learning for NLP. It begins with the core idea of how a word can be input into a neural network: the input embedding via a look-up table. The input embeddings, and the complementarity output embeddings obtained from the weights of the softmax layer, capture a representation of the words. The meaning of a word is determined by its usage, which is largely characterised by what words it co-occurs with. Word representation models include the well known skip-gram model which use these principles to derive high quality representations that can be reused as features in many tasks. The skip-gram model is based around predicting which words will be co-located. It is closely related to an iterative approximation to factorising the matrix of co-occurrence counts. This is similar to the approaches commonly used in older methods such as LSI and LDA. Word embedding models commonly use hierarchical softmax or negative sampling to speed-up the training and the evaluation.

Reminder: There is more to clustering than K-means
K-means is the most well-known clustering algorithm. But it is by no means the only one. K-means in particular is very vulnerable to getting stuck in local optima, compared to many other clustering algorithms. If one does use k-means, then make sure to run it multiple times and take the best result.

Chapter 4 extends the idea of representing words to representing senses. Most words have many meanings. It is thus impossible for a single representation to characterise the correct meaning in all contexts. Word senses can either be externally defined using a lexical resource, or discovered (induced) from the data. If the senses are externally defined, determining a good representation for them boils down to disambiguating a corpus to find the senses used, then creating a representation for them using the single sense word embedding methods. Inducing the senses from the corpus is more in-line with the general goal of not needing hand-engineered features, and there are many more methods for this.

The majority of word sense induction methods are either context-clustering-based, or co-location-prediction based. In both cases, the core idea is still that (like for word-embeddings) the surrounding words characterise what a particular word use means. In the *context clustering* methods, for each word the different contexts in which it occurs are represented and then clustered. Each cluster represents a word sense. In the *co-location prediction* approaches the word sense is treated as a hidden (latent) variable, which influencing the observable variables, i.e. the context words which it is co-occurring with. Classical probability methods can then be used, together with neural network methods, in order to uncover that latent variable that is the sense. These word sense induction methods give embeddings that can be used in much the same way as word embeddings.

Machine learning beyond neural networks
Though we have barely mentioned them in this book, there are plenty of other machine learning algorithms beyond just neural networks. Many of these have great advantages over neural networks, in terms of their performance with smaller amounts of data.
Some examples of some of the diversity of options (contrasted with neural networks) can be found at https://white.ucc.asn.au/2017/12/18/7-Binary-Classifier-Libraries-in-Julia.html.
It is often ideal to take word-embeddings (or other neural network derived representations) and use them as features, in a classifier such as a support vector machine (Cortes and Vapnik 1995), or a gradient boosting tree ensemble (Chen and Guestrin 2016).

Chapter 5 takes the idea of representing words to larger structures. There is a great diversity of methods to represent phrases, sentences, and documents. In three broad categories, we can consider weakly ordered models, sequential models, and structural models. The *weakly ordered models* are the most diverse category, being a catch-all term for a variety of methods that do not directly consider word order, including the neural embedding extensions of bag of words and bag of n-grams. The *sequential methods* are largely the application of the RNNs from Chap. 2, but also include some methods specifically for obtaining a representation. Finally, the *structural methods* allow the inputs to change the structure of the network, allowing it to process parse trees in a natural way.

All things considered, there are a great number of techniques for using natural language with machine learning. They all revolve around the core idea of a representation. Beyond that they range from simple to complex.

The methods used need to be sufficiently powerful to accomplish the task, but beyond this excess capacity is both a waste of training time, and a waste of developer effort. Spending more time to implement one of the more complicated models may not result in a final system that even works as well as a simple baseline. The ideal system is suitably simple. However, the identification of what it means for an implementation to be suitably simple takes surprising amounts of research. In part this can perhaps be attributed to our over-expectation of complexity in language. While language can be complex, a lot of it is surprisingly simple.

The systems we have discussed have been small and self-contained. They are suitable for use as a component in a larger system. If one looks at the workings of a digital assistant (such as the ones found on a smart phone), one will find many machine learning based subsystems: for speech-recognition, for intent detection, and for accomplishing subtasks within those. A complicated system such as Zara (Siddique et al. 2017) contains over a dozen separate machine learning based subsystems.

Zara, and defending herself from bad influences
One of the most interesting subsystems in Zara (Siddique et al. 2017), is the use of a module to detect abusive language, racism and sexism. It is a well-known problem that when allowing a learning system to learn from the open-web, such systems can pick-up bad habits (https://www.theverge.com/2016/3/24/ 11297050). The use of such an abuse detection module allows a system to be protected from such things.

This can be related to our earlier comments: with regards to the issues with WordNet's sexist definitions, and the issues that word embeddings can learn the biases representations from biases texts discussed by Bolukbasi et al. (2016) and Caliskan et al. (2017)

There is another approach, processing huge corpora of texts, with end-to-end deep learning systems. For example using decades of user support logs to train question answering systems. Even this is based on many of the same principles we have discussed in this book. Though, at this scale other options open up, like char-RNNs which function on a character scale rather than the word-scale that we have considered here. With enough data, all the linguistic modelling concerns can be learnt into the neural network. The data and computational requirements place such things out of reach of many developers.

Most tasks do not have this scale of data. They may be small data, a few hundred examples. They may be largish data, with a few million examples. The more data you have, the more complex a model you can train, and thus the more difficult a problem you can solve. But good representations can help us when we do not have enough data. By creating word embeddings (Chap. 3), word sense embeddings (Chap. 4), and larger embeddings (Chap. 5), we can leverage other larger data sources. The embeddings created using systems that are trained on other data, to perform other tasks, capture information to accomplish those tasks. We can then take those information dense embeddings, and apply them to our own tasks.

In conclusion, this book has covered popular deep neural networks for language processing. The networks have been comprehensively reviewed and explained from a practical perspective. Language modelling, vector representations, and challenging tasks such as WSD, and sentence embeddings were investigated to illustrate the use of these networks. This book shares a lot of valuable advice obtained from first-hand experience in implementing these networks. We would treat this book as a solid introduction to one of the most exciting new areas of natural language processing and computational linguistics.

References

Bolukbasi, Tolga, Kai-Wei Chang, James Y. Zou, Venkatesh Saligrama, and Adam T. Kalai. 2016. Man is to computer programmer as woman is to homemaker? Debiasing word embeddings. In *Advances in neural information processing systems*, 4349–4357.

Caliskan, Aylin, Joanna J. Bryson, and Arvind Narayanan. 2017. Semantics derived automatically from language corpora contain human-like biases. *Science 356* (6334), 183–186. ISSN: 0036-8075, https://doi.org/10.1126/science.aal4230, http://science.sciencemag.org/content/356/6334/183.full.pdf.

Chen, Tianqi and Carlos Guestrin. 2016. Xgboost: A scalable tree boosting system. In *Proceedings of the 22nd ACM SIGKDD international conference on knowledge discovery and data mining*, 785–794. ACM.

Cortes, Corinna and Vladimir Vapnik. 1995. Support-vector networks. *Machine Learning* 20 (3): 273–297. ISSN: 1573-0565, https://doi.org/10.1007/BF00994018.

Siddique, Farhad Bin, Onno Kampman, Yang Yang, Anik Dey, and Pascale Fung. 2017. Zara returns: Improved personality induction and adaptation by an empathetic virtual agent. In *Proceedings of ACL 2017, system demonstrations*, 121–126.

Index

© Springer Nature Singapore Pte Ltd. 2019
L. White et al., *Neural Representations of Natural Language*,
Studies in Computational Intelligence 783,
https://doi.org/10.1007/978-981-13-0062-2